Water Pollution XV

WITPRESS

WIT Press publishes leading books in Science and Technology.
Visit our website for the current list of titles.
www.witpress.com

WITeLibrary

Home of the Transactions of the Wessex Institute.
Papers contained in this volume are archived in the WIT eLibrary in volume 242 of WIT
Transactions on Ecology and the Environment (ISSN 1743-3541).
The WIT eLibrary provides the international scientific community with immediate and
permanent access to individual papers presented at WIT conferences.
Visit the WIT eLibrary at www.witpress.com.

FIFTEENTH INTERNATIONAL CONFERENCE ON
MONITORING, MODELLING AND MANAGEMENT OF
WATER POLLUTION

Water Pollution 2020

CONFERENCE CHAIRMEN

Stefano Mambretti
Polytechnic of Milan, Italy
Member of WIT Board of Directors

José Sergio Palencia Jiménez
Polytechnic University of Valencia, Spain

HONORARY CHAIRMAN

Francisco José Mora Mas
Polytechnic University of Valencia, Spain

INTERNATIONAL SCIENTIFIC ADVISORY COMMITTEE

ORGANISED BY
Wessex Institute, UK
Polytechnic University of Valencia, Spain

SPONSORED BY
WIT Transactions on Ecology and the Environment
International Journal of Environmental Impacts

WIT Transactions

Wessex Institute
Ashurst Lodge, Ashurst
Southampton SO40 7AA, UK

K. Dorow Pacific Northwest National Laboratory, USA

W. Dover University College London, UK

C. Dowlen South Bank University, UK

J. P. du Plessis University of Stellenbosch, South Africa

R. Duffell University of Hertfordshire, UK

A. Ebel University of Cologne, Germany

V. Echarri University of Alicante, Spain

K. M. Elawadly Alexandria University, Egypt

D. Elms University of Canterbury, New Zealand

M. E. M El-Sayed Kettering University, USA

D. M. Elsom Oxford Brookes University, UK

F. Erdogan Lehigh University, USA

J. W. Everett Rowan University, USA

M. Faghri University of Rhode Island, USA

R. A. Falconer Cardiff University, UK

M. N. Fardis University of Patras, Greece

A. Fayvisovich Admiral Ushakov Maritime State University, Russia

H. J. S. Fernando Arizona State University, USA

W. F. Florez-Escobar Universidad Pontifica Bolivariana, South America

E. M. M. Fonseca Instituto Politécnico do Porto, Instituto Superior de Engenharia do Porto, Portugal

D. M. Fraser University of Cape Town, South Africa

G. Gambolati Universita di Padova, Italy

C. J. Gantes National Technical University of Athens, Greece

L. Gaul Universitat Stuttgart, Germany

N. Georgantzis Universitat Jaume I, Spain

L. M. C. Godinho University of Coimbra, Portugal

F. Gomez Universidad Politecnica de Valencia, Spain

A. Gonzales Aviles University of Alicante, Spain

D. Goulias University of Maryland, USA

K. G. Goulias Pennsylvania State University, USA

W. E. Grant Texas A & M University, USA

S. Grilli University of Rhode Island, USA

R. H. J. Grimshaw Loughborough University, UK

D. Gross Technische Hochschule Darmstadt, Germany

R. Grundmann Technische Universitat Dresden, Germany

O. T. Gudmestad University of Stavanger, Norway

R. C. Gupta National University of Singapore, Singapore

J. M. Hale University of Newcastle, UK

K. Hameyer Katholieke Universiteit Leuven, Belgium

C. Hanke Danish Technical University, Denmark

Y. Hayashi Nagoya University, Japan

L. Haydock Newage International Limited, UK

A. H. Hendrickx Free University of Brussels, Belgium

C. Herman John Hopkins University, USA

I. Hideaki Nagoya University, Japan

W. F. Huebner Southwest Research Institute, USA

M. Y. Hussaini Florida State University, USA

W. Hutchinson Edith Cowan University, Australia

T. H. Hyde University of Nottingham, UK

M. Iguchi Science University of Tokyo, Japan

L. Int Panis VITO Expertisecentrum IMS, Belgium

N. Ishikawa National Defence Academy, Japan

H. Itoh University of Nagoya, Japan

W. Jager Technical University of Dresden, Germany

Y. Jaluria Rutgers University, USA

D. R. H. Jones University of Cambridge, UK

N. Jones University of Liverpool, UK

D. Kaliampakos National Technical University of Athens, Greece

D. L. Karabalis University of Patras, Greece

A. Karageorghis University of Cyprus

T. Katayama Doshisha University, Japan

K. L. Katsifarakis Aristotle University of Thessaloniki, Greece

E. Kausel Massachusetts Institute of Technology, USA

H. Kawashima The University of Tokyo, Japan

B. A. Kazimee Washington State University, USA

F. Khoshnaw Koya University, Iraq

S. Kim University of Wisconsin-Madison, USA

D. Kirkland Nicholas Grimshaw & Partners Ltd, UK

E. Kita Nagoya University, Japan

A. S. Kobayashi University of Washington, USA

D. Koga Saga University, Japan

S. Kotake University of Tokyo, Japan

Water Pollution XV

Editors

Stefano Mambretti
Polytechnic of Milan, Italy
Member of WIT Board of Directors

José Sergio Palencia Jiménez
Polytechnic University of Valencia, Spain

WITPRESS Southampton, Boston

Editors:

Stefano Mambretti
Polytechnic of Milan, Italy
Member of WIT Board of Directors

José Sergio Palencia Jiménez
Polytechnic University of Valencia, Spain

Published by

WIT Press
Ashurst Lodge, Ashurst, Southampton, SO40 7AA, UK
Tel: 44 (0) 238 029 3223; Fax: 44 (0) 238 029 2853
E-Mail: witpress@witpress.com
http://www.witpress.com

For USA, Canada and Mexico

Computational Mechanics International Inc
25 Bridge Street, Billerica, MA 01821, USA
Tel: 978 667 5841; Fax: 978 667 7582
E-Mail: infousa@witpress.com
http://www.witpress.com

British Library Cataloguing-in-Publication Data

A Catalogue record for this book is available
from the British Library

ISBN: 978-1-78466-383-4
eISBN: 978-1-78466-384-1
ISSN: 1746-448X (print)
ISSN: 1743-3541 (on-line)

The texts of the papers in this volume were set individually by the authors or under their supervision. Only minor corrections to the text may have been carried out by the publisher.

Preface

This volume contains the papers presented at the 15th International Conference on Monitoring, Modelling and Management of Water Pollution, scheduled in Valencia but held on line due to the Coronavirus pandemic, organised by the Wessex Institute together with the Polytechnic University of Valencia, Spain. The Water Pollution Conference started in Southampton, UK in 1991 and has been held in Milan, Italy (1993); Porto Carras, Greece (1995); Lake Bled, Slovenia (1997); Lemnos, Greece (1999); Rhodes, Greece (2001); Cadiz, Spain (2003); Bologna, Italy (2006); Alicante, Spain (2008); Bucharest, Romania (2010); The New Forest, UK, home of the Wessex Institute (2012); the Algarve (2014), Venice (2016) and A Coruna (2018).

The availability of unlimited water resources cannot any longer be taken for granted as the needs of a growing world population, demanding better standards of living, continues to increase. Prominent among those problems is water quality that, due to the increase of pollutant loads discharged into natural water bodies, requires better tools for assessment and the formation of a framework for regulation and control. This framework needs to be based on scientific results that relate pollutant discharge with changes in water quality. The results of these studies allow the industry to apply more efficient methods of controlling and treating waste loads, and water authorities to enforce appropriate regulations regarding this matter.

Contamination of water resources comes from very different sources, including industrial, agricultural and residential users. This diversity of usage results in the need to understand better the complex physio-chemical process involved. Moreover, environmental problems are essentially interdisciplinary. Engineers and scientists working in this field must be familiar with a wide range of issues including the physical processes of mixing and dilution, chemical and biological processes, mathematical modelling, data acquisition and measurement, to name but a few.

Furthermore, water quality can have dramatic effects on human health, not only due to heavy metals and other well-known agents, but also a wide range of emerging chemical and pharmaceutical products whose effects are poorly understood. In view of the scarcity of available data, it is important that experiences are shared on an international basis. Thus, a continuous exchange of information between scientists from different countries is essential.

Moreover, in the last years the topic of the Contaminants of emerging concern (CECs) forcefully arose, due to pharmaceuticals, cyanotoxins, personal care products, nanoparticles, and flame retardants, among others. These classifications are constantly changing as new contaminants (or effects) are discovered and emerging contaminants from past years become less of a priority. These contaminants can generally be categorized as truly "new" contaminants that have only recently been discovered and researched, contaminants that were known about but their environmental effects were not fully understood, or "old" contaminants that have new information arising regarding their risks. The topic is presently of paramount importance in the field of water pollution.

The papers in this book make a significant contribution to the solution of some of these issues.

These papers, like others presented at Wessex Institute conferences, are referenced by CrossRef and appear regularly in suitable reviews, publications and databases, including referencing and abstracting services. They are also archived online in the WIT eLibrary (http://www.witpress.com/elibrary) where they are permanently available in Open Access format to the international scientific community.

The Editors would like to thank the authors for their contributions, as well as the member of the International Scientific Advisory Community of the Conference for their invaluable help in reviewing the papers.

The Editors, 2020

Contents

SECTION 1
WATER CONTAMINATION

HYDROCHEMICAL PARAMETERS IN A PORTION OF THE PARAÍBA DO SUL RIVER HYDROGRAPHIC BASIN, SÃO JOSÉ DOS CAMPOS CITY, SÃO PAULO STATE, BRAZIL

ISABELLA G. LEE & DANIEL M. BONOTTO
Departamento de Geologia, Instituto de Geociências e Ciências Exatas – UNESP, Rio Claro, Brazil

ABSTRACT

This paper describes a hydrochemical study held in a portion of the Paraíba do Sul river hydrographic basin that is located in São José dos Campos city at eastern of São Paulo State, Brazil. It comprises Lambari river, a tributary of the Paraíba do Sul river, which crosses the installations of the Henrique Lage Oil Refinery (REVAP). Here is reported the hydrochemical parameters of the Lambari river waters that drain the refinery in order to identify possible impacts on the water resources due to the presence of the refinery. All samples analyzed (32) exhibited a similar hydrochemical pattern, except sample IL-02, which presented different values for most of the parameters analyzed. The samples exhibited pH values between 6.16 and 7.41 and Eh values between +260 and +324 mV, except for one outlier value for sample IL-02 (−170 mV), which is reduced, according to Eh-pH diagram. The following mean values were found for the parameters analyzed in this study, disregarding the outlier sample IL-02: temperature = 26.4°C; Electrical Conductivity (EC) = 108.03 µS/cm; alkalinity = 32 mg/L; turbidity = 6.8 FTU; Total Dissolved Solids (TDS) = 62.9 mg/L; Ca^{2+} = 2.6 mg/L; Mg^{2+} = 1.04 mg/L; Na^+ = 13.6 mg/L; K^+ = 2.8 mg/L; Cl^- = 8.2 mg/L; NO_3^- = 1.2 mg/L; SO_4^{2-} = 1.5 mg/L. Based on the mean hydrochemical composition, the waters were classified as sodium bicarbonate. An analysis was performed to determine surfactants and tannin-lignin in the sample IL-02, resulting in values of 0.302 mg/L and 2.6 mg/L, respectively. It is believed that this sample has an excessive contribution of domestic effluents, since it has high levels of surfactants and its collection point is located close to a sewage pipe.
Keywords: surface waters, hydrochemical parameters, oil refinery.

1 INTRODUCTION

Surface waters are very important for the society maintenance because they represent one of the main sources of water supply in the planet. Thus, hydrochemical studies focusing the surface waters quality are necessary for proper planning and management of the water resources in a region. Watercourses in metropolitan regions are impacted by anthropogenic activities taking place in numerous drainages, which are related to both constant population growth and industrial practices, thus, requiring greater exploitation of the water resources and causing the water degradation due to the inputs of domestic and industrial effluents. This paper describes a hydrochemical study held in a portion of the Paraíba do Sul river hydrographic basin. The study area is located in the Paraíba valley region that is an important economic axis between São Paulo and Rio de Janeiro in Brazil, which possesses a great demand for water resources because comprises large and medium-sized cities, such as Jacareí, Taubaté, Aparecida do Norte and São José dos Campos at São Paulo State.

More specifically, this study was held at São José dos Campos city, located in the east of São Paulo State, focusing the Henrique Lage Refinery (REVAP) area, one of the most remarkable oil refineries from Petrobras in Brazil (Fig. 1). The site is located in a portion of the Paraíba do Sul river hydrographic basin, comprising Lambari river, a tributary of the Paraíba do Sul river, which crosses the installations of the refinery. São José dos Campos city is home of important companies and research centers in the country. In addition, it is an

WIT Transactions on Ecology and the Environment, Vol 242, © 2020 WIT Press
www.witpress.com, ISSN 1743-3541 (on-line)
doi:10.2495/WP200011

important technopole of warlike and metallurgical materials, as well the headquarters of the largest aerospace complex in Latin America.

Because of the increasing use of water resources and the difficulty of maintaining and recovering them in cases of contamination, the understanding of the physical, physicochemical and chemical characteristics of the waters is essential for their proper management. Thus, due to the proximity of the oil refinery to the study area, the hydrochemical analysis of surface water in the region are essential for their appropriate use. This research aims to describe some hydrochemical parameters of the Lambari river waters that are close to the refinery in order to identify possible changes in their quality associated to the presence of the refinery.

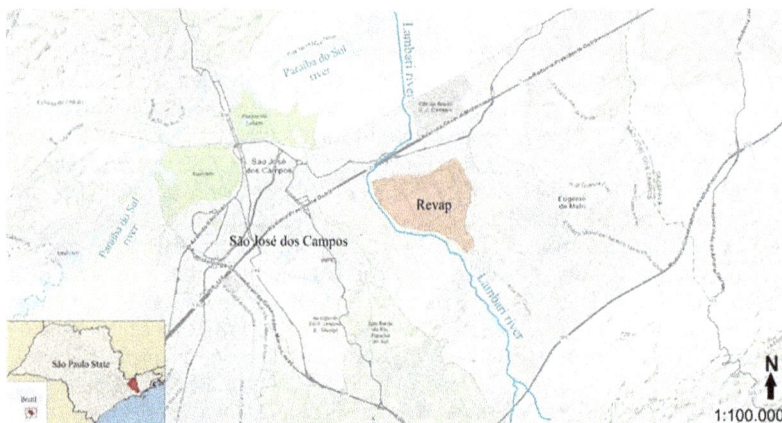

Figure 1: Location of the study area.

2 STUDY AREA

The study area is inserted in the Paraíba do Sul river hydrographic basin which covers an area of approximately 57,000 km², and extends to three states in the southeastern region of Brazil: São Paulo, Minas Gerais and Rio de Janeiro. In the state of São Paulo, the Paraíba do Sul river hydrographic basin covers several cities, including Cruzeiro, Guaratinguetá, Jacareí, Lorena, São José dos Campos, Taubaté and Aparecida do Norte. The main water body in this hydrographic basin is the Paraíba do Sul river, which is formed by the confluence of the Paraitinga and Paraibuna rivers. The main tributaries are the Jaguari, Paraibuna, Pomba and Muriaé rivers, on the left bank, and the Piraí, Piabanha and Dois Rios rivers, on the right bank. It flows into the Atlantic Ocean, discharging at São João da Barra, Rio de Janeiro State [1].

The Paraíba do Sul river basin is divided into four sub-basins: Low Paraíba do Sul basin, Muriaé/Pomba basin, Paraibuna basin, and High Paraíba do Sul Basin. The Lambari river, focused in this study, is a tributary of Paraíba do Sul river and is located in the High Paraíba do Sul sub-basin, more accurately at São José dos Campos city. It flows from southeast to northwest of São José dos Campos city, crossing the Henrique Lage Refinery (REVAP) (Fig. 1). The major access to the site is by Presidente Dutra highway (BR-116) that links São Paulo to Rio de Janeiro cities.

The study area is inserted in the geological context of Taubaté basin, which is a rift-type basin characterized by continental and syntectonic sedimentation associated to fluvial and lacustrine environments [2]–[4]. Geomorphologically, it is situated at Paraíba do Sul river valley that is a lowered surface embedded in Precambrian rocks limited by Mantiqueira mountain range at northwestern and Mar mountain range at southeastern [5]. The sedimentation comprised a phase syntectonic to the rift (deposition of sediments from Taubaté Group that include the formations Resende, Tremembé and São Paulo) and a phase after the dystrophic tectonics (deposition of Pindamonhangaba Formation and alluvial/colluvial sediments) [4].

The climate is warm and temperate at São José dos Campos region, with a significant rainfall along the year. It is classified as Cfa according to the Köppen and Geiger classification [6] as exhibits an average temperature of 19.4°C and mean annual rainfall of 1269 mm. The remaining vegetation consists on the Atlantic Forest that dominates the slopes of Mantiqueira mountain range and banks of Paraíba do Sul river.

3 MATERIALS AND METHODS

This study started with fieldwork for the samples collection and involved *in situ* tests and laboratorial analysis. The collection of water samples from the Lambari river was conducted in late October and early November 2019 at 30 previously defined points that were regularly distributed throughout the perimeter of interest, with an interval of approximately 400 m between them (Fig. 2). In addition, two samples were also collected at Paraíba do Sul river in points situated upstream and downstream of the Lambari river discharge. The water sampling was performed according to location, access and functionality. The coordinates of all sampling points were recorded during the fieldwork, as well as their collection time.

Figure 2: Location of the 32 samples collection points.

Polyethylene terephthalate containers duly identified and numbered were used for sampling, which were previously washed and rinsed at the collection time with the water sampled, in order to avoid contamination. It was initially recovered about 5L of water from each point and the temperature readings were done with a precision reading thermometer, as quickly as possible, in order to avoid changes in its value. For the analysis of the physicochemical parameters, the samples were stored in 2L-volume polyethylene flasks. After checking the hydrogen potential (pH), oxidation-reduction potential (Eh), electrical conductivity (EC), alkalinity and turbidity, the samples were carefully handled and taken to LABIDRO-Isotopes and Hydrochemistry Laboratory, IGCE-UNESP – Rio Claro Campus, for hydrochemical analysis of sodium (Na^+), potassium (K^+), calcium (Ca^{2+}), magnesium (Mg^{2+}), chloride (Cl^-), sulfate (SO_4^{2-}), nitrate (NO_3^-), surfactants, and tannin-lignin.

The pH was measured with a portable Digimed digital equipment, which was coupled to a combined glass electrode model KASVI. The determination of Eh was performed with the same device, however, it was coupled to a combined Digimed platinum electrode, model 12L12069, which was calibrated with a Zobell solution, according to the procedure described by Bonotto [7].

The electrical conductivity (EC) was measured with one Analion digital equipment, model C-702. Alkalinity was measured by titration with 0.02 N sulfuric acid by the Hach method 8221 – Buret Titration whose readings are in the concentration range between 0 and 500 ± 0.2 mg/L [8]. Turbidity was determined by the Hach method 8237 – Absorptometric, using the Hach spectrophotometer, model DR/2000 [8].

The determination of sodium and potassium was made by the flame atomic emission spectrometry method, through the Benfer flame photometer, model BFC-300. The calcium and magnesium data were obtained by the Hach method 8030 – Calmagite Colorimetric, chloride by the Hach method 8113 – Mercuric Thiocyanate, and sulfate by the Hach method 8051 – SulfaVer 4, all them using the Hach DR/2000 spectrophotometer [8]. Nitrate was measured by the Hach method 10020 – Chromotropic Acid in a Hach DR/2700 spectrophotometer [8].

The summation of the concentration of the main cations (Na^+, K^+, Ca^{2+}, Mg^{2+}) and anions (Cl^-, SO_4^{2-}, NO_3^-, HCO_3^-) in the analyzed waters allowed determine the Total Dissolved Solids concentration (TDS 1). In addition, in order to check such results, it was estimated another TDS value (TDS 2) of the samples through the theoretical relationship between TDS and EC in natural waters as proposed by Hem [9]:

$$\text{TDS 2 (in mg/L)} = \text{EC (in } \mu S/cm \text{ at } 25°C) \times A, \qquad (1)$$

where A is a factor ranging from 0.5 to 1.0. However, usually A is between 0.55 and 0.75, except for waters exhibiting unusual composition. In this study, it was adopted A = 0.65 that corresponds to the average of the most common A values.

Surfactants and tannin-lignin analysis were also performed in order to obtain an explanation for the fact that one sample presented outliers values of Eh, EC, TDS, alkalinity, turbidity, calcium, sodium, potassium, chloride, nitrate and sulfate. Surfactants data often indicate the presence of domestic effluents in the waters such as cleaning products (detergents) and cosmetic products (shampoos), whilst tannin-lignin data identify possible effects of plant decomposition in the waters. For comparison purposes, the analysis was made for the outlier sample (IL-02) and another sample (IL-39) that exhibited the second highest measured EC value. The surfactants were measured using the Hach method 8028 – Crystal Violet [8], whereas the tannin-lignin data were obtained using the Hach method 8193 – Tyrosine [8]. In both cases, the Hach DR/2000 spectrophotometer model was used for the readings. In order to characterize the hydrochemical facies of the waters, a free software (Qualigraf, v. 1.17) was used, which was developed by FUNCEME – Ceará Meteorology and Water Foundation.

4 RESULTS AND DISCUSSION

Table 1 shows the results obtained in this study. The temperature measured *in situ* varied little, with values ranging from 24.6°C to 32.8°C. The pH values indicated that the analyzed waters are basically neutral, with little variation in pH, ranging from 6.16 to 7.41. The Eh values are between +260 and +324 mV, except for the outlier value of the sample IL-02 (−170 mV). Based on the pH and Eh data, the waters of the Lambari river can be classified as slightly oxidizing, except the sample IL-02 that is reduced as shown in the Eh-pH diagram (Fig. 3) that characterizes the type of environment where the water is flowing.

Table 1:   Minimum (Min.), maximum (Max.) and mean values obtained for the parameters analyzed in the 32 water samples.

Parameter	Unit	Min.	Max.	Mean
Temperature	°C	24.6	32.8	26.4
Eh	mV	−170	324	299
pH	–	6.16	7.41	6.89
EC	µS/cm	72.6	560	108.03
TDS 1	mg/L	42.27	284.45	62.93
TDS 2	mg/L	47.19	364	70.22
ΔTDS[1]	mg/L	0.19	79.55	9.89
Turbidity	FTU	1	120	6.84
Sodium	mg/L	7.36	52.21	13.57
Potassium	mg/L	0.59	10.32	2.85
Calcium	mg/L	0.03	11.40	2.6
Magnesium	mg/L	0.56	1.66	1.04
Alkalinity	mg/L	22	108	32
Chloride	mg/L	1.90	48.80	8.16
Nitrate	mg/L	0.40	25.80	1.25
Sulfate	mg/L	<1	27	1.52

[1]ΔTDS = |TDS 2 – TDS 1|.

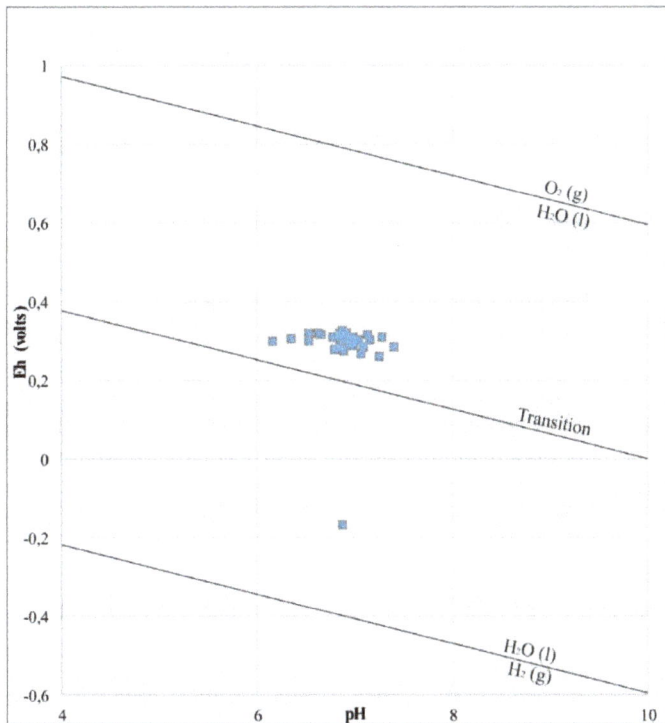

Figure 3:   Data obtained for the waters analyzed in this study inserted in an Eh-pH diagram.

The estimated TDS concentration (TDS 2) yielded satisfactory results, correlating significantly with TDS 1 and providing a difference relative to TDS 1 (ΔTDS) that corresponded to a mean value of 9.9 mg/L (Table 1). Thus, the hydrochemical data have been discussed here considering the TDS 1 data.

The EC of the water samples ranged from 72.6 to 214 μS/cm, the alkalinity between 22 and 64 mg/L of CaCO3, the turbidity between 1 and 42 FTU, and the TDS 1 from 42.27 to 115.1 mg/L, except the sample IL-02, which showed values of 560 μS/cm (EC), 108 mg/L (alkalinity), 120 FTU (turbidity) and 284.45 mg/L (TDS 1).

The TDS 1-EC graph is shown in Fig. 4, showing a strong linear correlation (Pearson's coefficient r = 0.98). For turbidity, it was found that 10 samples (31%) exhibited levels exceeding the maximum value allowed by the Brazilian Ministry of Health for drinking water corresponding to 5 FTU [10]. For TDS, all values are within the drinking requirements of the Brazilian Ministry of Health corresponding to 1000 mg/L [10].

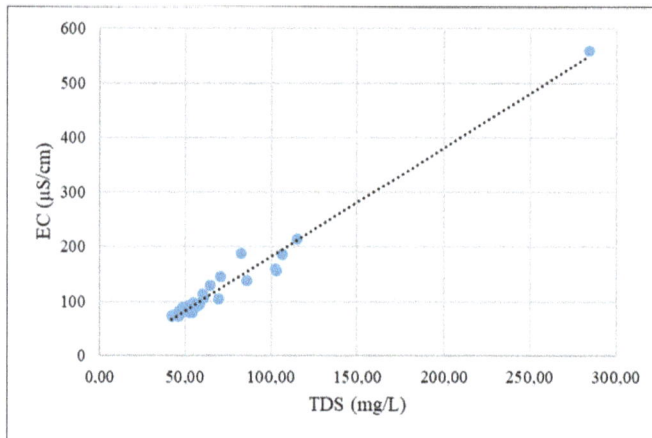

Figure 4: The TDS 1-EC graph obtained for all samples analyzed.

The graphs in Figs 5 and 6 show the relationships among alkalinity, EC and TDS 1. They indicate good data correlation, with Pearson coefficients of r = 0.89 (Fig. 5) and r = 0.90 (Fig. 6), characterizing high linear correlation among these parameters. The linear dependence between alkalinity and TDS 1 is consequence of the TDS 1-EC relationship. Bicarbonate (as indicated by alkalinity) often takes a major role in the chemical composition of the surface waters and it is a major constituent of the TDS 1 in the studied waters, offering a significant correlation with this parameter. It was also carried out a statistical analysis disregarding the outlier sample IL-02, but even in this case, the linear dependence still exists as the Pearson correlation coefficient corresponded to r = 0.66 (alkalinity *vs.* EC) and r = 0.70 (alkalinity *vs.* TDS 1).

The pH, in turn, showed practically no linear dependence with EC and alkalinity (Pearson's correlation coefficient of –0.19 and –0.06, respectively). The Eh variation in relation to the pH values for the study area was analyzed disregarding the outlier sample IL-02 (Fig. 7). The relationship between these two parameters is inversely proportional as often recognized in the literature. In our case, there is a slight trend of this inverse relationship as indicated by the Pearson correlation coefficient r = –0.31 (Fig. 7).

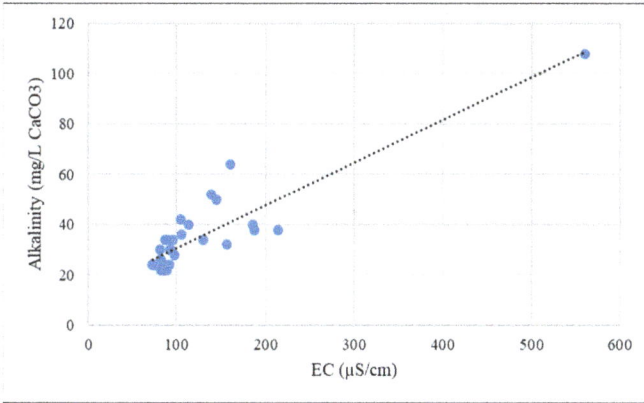

Figure 5: The alkalinity-EC graph obtained for all water samples analyzed.

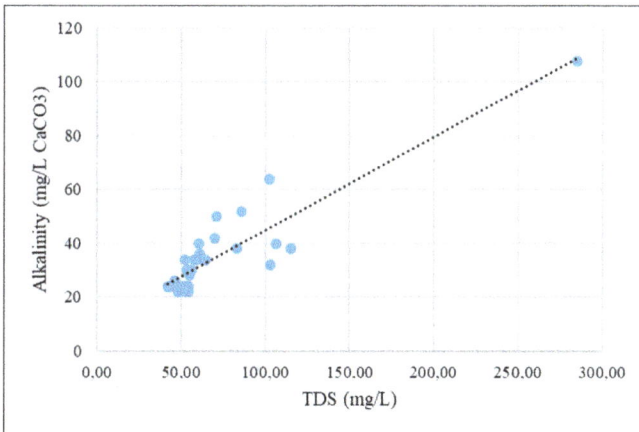

Figure 6: The alkalinity-TDS 1 graph obtained for all water samples analyzed.

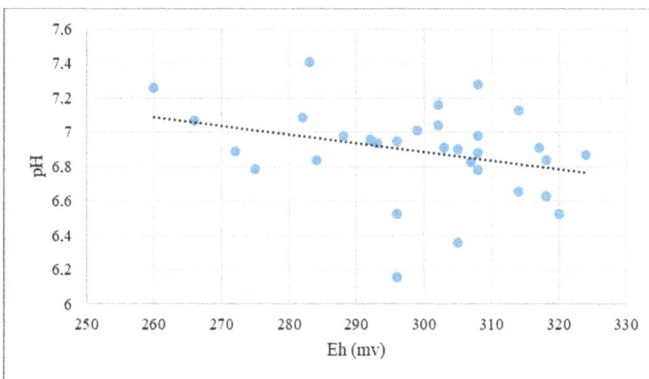

Figure 7: The pH-Eh graph obtained disregarding the outlier sample IL-02.

When analyzing the turbidity in relation to the TDS 1 concentration, a positive linear dependence was found (Pearson's correlation coefficient $r = 0.90$) (Fig. 8) that is expected once the turbidity of a water body is directly associated with the levels of dissolved and suspended solids. These parameters were again submitted to the statistical analysis, disregarding the influence of the outlier sample and, in this case, the Pearson correlation corresponded to $r = 0.43$, still showing a moderate linear dependence between these parameters.

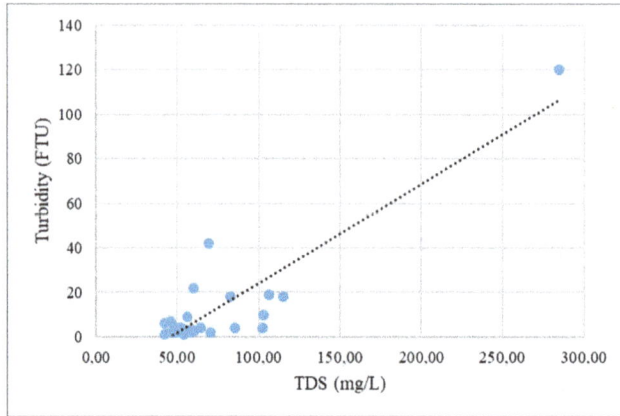

Figure 8: The turbidity-TDS 1 graph obtained for all water samples analyzed.

The calcium concentration was between 0.03 and 5.24 mg/L, except for the sample IL-02 that corresponded to 11.4 mg/L. For magnesium, the range was 0.56–1.66 mg/L. The Brazilian Ministry of Health [10] allows up to 500 mg/L of Ca and Mg in drinking water that is a value well above of those found in the analyzed waters.

The maximum sodium concentration in drinking water allowed by the Brazilian Ministry of Health is 200 mg/L [10] that is above the values found here (7.36 to 27.83 mg/L), even for sample IL-02 (52.21 mg/L). For potassium, the range was 0.59–6.99 mg/L, except for sample IL-02 (10.32 mg/L).

The maximum nitrate concentration in drinking water allowed by the Brazilian Ministry of Health is 10 mg/L [10] that is above the range 0.4–2.4 mg/L found here, except for sample IL-02 that surpassed it (25.8 mg/L), which may refer to contamination by septic tanks and sewage effluents.

The chloride levels were between 1.9 and 36.2 mg/L, except for sample IL-02 (48.8 mg/L), but all these values are lower than the guideline reference value of 250 mg/L in drinking water, according to the Brazilian Ministry of Health [10].

The maximum sulfate concentration in drinking water allowed by the Brazilian Ministry of Health is 250 mg/L [10] that is above the values found here (<1 to 9 mg/L), even for sample IL-02 (27 mg/L).

In addition to the alkalinity-EC relationship, sodium and chloride also correlated with the EC. For Na^+, the Pearson's coefficient was $r = 0.91$ (Fig. 9), whilst it was $r = 0.84$ for Cl^- (Fig. 10). The correlation was significant even disregarding the outlier values of the sample IL-02 as it was found $r = 0.82$ for Na^+ (Fig. 11) and $r = 0.78$ for Cl^- (Fig. 12).

The dominant hydrochemical composition of surface waters often reported in the literature is calcium-bicarbonated but in the study area it tends to be sodium-bicarbonated (Table 1). The strong significant correlations between EC and alkalinity (bicarbonate)

(Fig. 5), as well between EC and sodium (Fig. 9) confirm such trend that is more accurately viewed in a Piper diagram [11] as shown in Fig. 13.

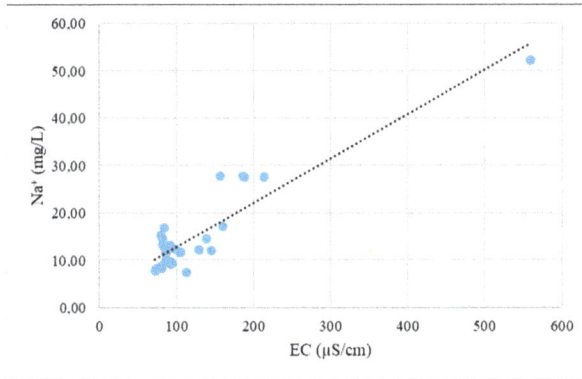

Figure 9: The Na-EC graph obtained from all samples analyzed.

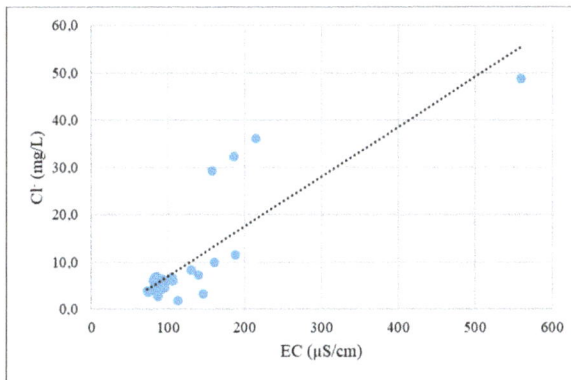

Figure 10: The Cl-EC graph obtained from all samples analyzed.

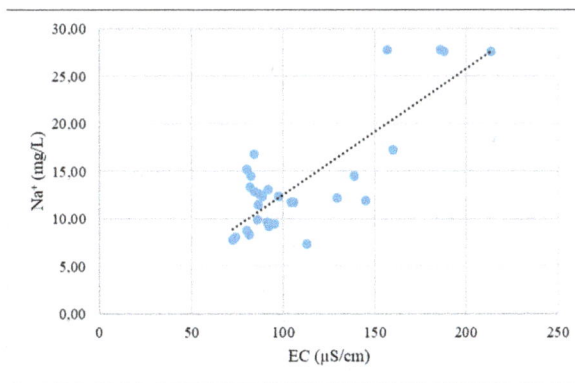

Figure 11: The Na-EC graph obtained disregarding the outlier sample IL-02.

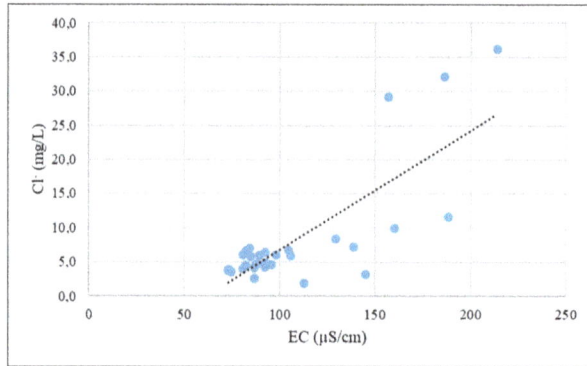

Figure 12: The Cl-EC graph obtained disregarding the outlier sample IL-02.

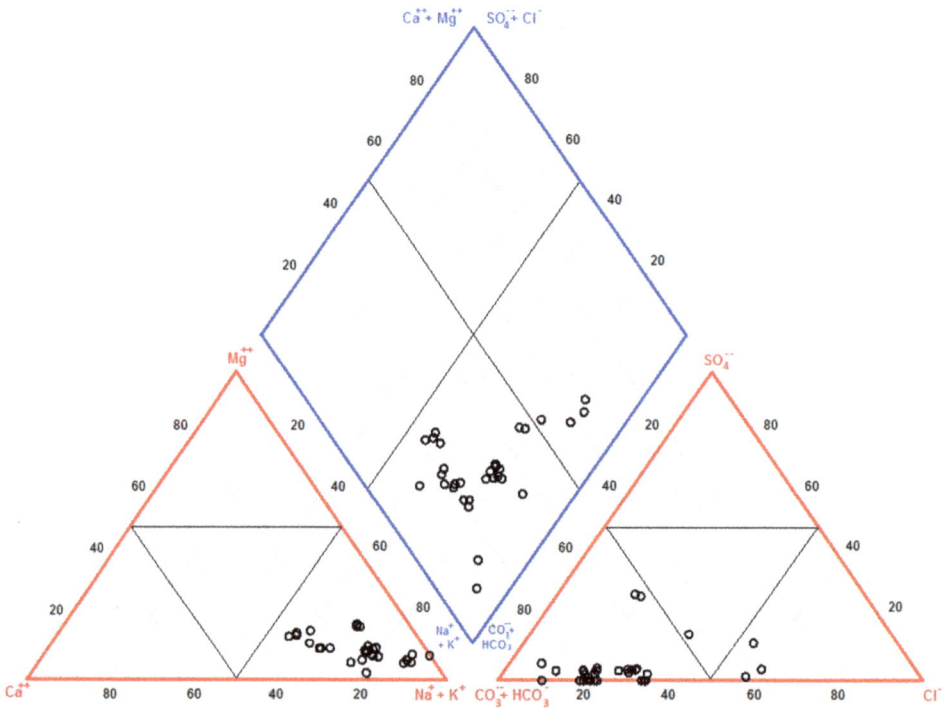

Figure 13: Chemical data of the analyzed samples (circles) plotted on a Piper diagram [11].

The insertion of the chemical data for the 32 samples analyzed in the Piper diagram [11] (Fig. 13) indicated that all them are classified as sodium-type waters, considering the dissolved cations. In terms of the dissolved anions, 88% are inserted in the field of the bicarbonate-type waters and 9% in the field of the chloride-type waters; the remaining 3% corresponds to the outlier sample IL-02 that is inserted in the field of the mixed-type waters.

The analysis to determine the values of surfactants for samples IL-02 and IL-39 indicated, respectively, 0.302 mg/L and 0.011 mg/L, suggesting an accentuated presence of surfactants in the outlier sample IL-02. The high value found in this sample is close to the maximum allowed by the Brazilian Ministry of Health (0.5 mg/L) [10] and, along with its high concentration of chloride (48.8 mg/L) and nitrate (25.8 mg/L), suggests contamination by domestic effluents, such as cleaning products, cosmetics and sewage effluents. It is likely that the outliers of some of the physicochemical parameters analyzed in the sample IL-02 are related to these contaminants inputs. The analysis made to determine the tannin-lignin levels in the water samples IL-02 and IL-39 indicated values of 2.6 mg/L and 0.8 mg/L, respectively. Thus, it is observed that the sample IL-02 has a greater influence in its hydrochemistry of plants decomposition, although not very significant.

5 CONCLUSION

The results of the present investigation focusing the waters of the Lambari river at São José dos Campos city, São Paulo State, Brazil, took into account parameters that allowed a helpful assessment of their quality, identifying possible environmental impacts occurring there. The physicochemical parameters analyzed in this paper confirmed that the Henrique Lage Oil Refinery (REVAP) did not influence their characteristics. However, additional studies are necessary to confirm such preliminary scenario, taking into account other parameters related to the water quality, for instance, minor and traces inorganic constituents, as well as organic compounds (BTEX – Benzene, Toluene, Ethyl-benzene and Xylenes). Sodium, chloride and alkalinity (bicarbonate) are the dominant constituents in the Lambari River waters, showing excellent linear correlation with the electrical conductivity (EC). The Piper diagram indicated that they are mostly sodium bicarbonate (88% of the samples), but also sodium chloride (9% of the samples). One outlier sample (IL-02) was classified as sodium mixed, which presented discrepant values for most of the parameters analyzed, except for pH and Mg. In that monitoring point, the composition could be related to an excessive contribution of domestic effluents as evidenced by its high surfactants level and also by sewage effluents as indicated by its higher concentration of chloride and nitrate. This is corroborated by the fact that the collection point of this sample is located close to a sewage pipe.

REFERENCES

[1]   Amorim, D.D. & Ferreira, M.E., Um estudo sobre a qualidade das águas do Rio Paraíba do Sul no Vale do Paraíba do Sul no período de 1978 a 1994. Presented at *XIII Brazilian Symposium on Water Resources*, Belo Horizonte, MG, 2000.

[2]   Appi, C.J., Freitas, E.L. & Castro, J.C., Faciologia e estratigrafia da Bacia de Taubaté. Technical report, CENPES/ Petrobras, 1986.

[3]   Chang, H.K. et al., Geologia da Bacia de Taubaté. Presented at *Southeastern Geology Symposium*, Rio de Janeiro, RJ, p. 10, 1989.

[4]   Riccomini, C., O rift continental do sudeste do Brasil. PhD thesis, University of São Paulo, São Paulo, 1989.

[5]   Ross, J. & Moroz, I., Mapa Geomorfológico do Estado de São Paulo. *Journal of Geography Department*, pp. 41–58, 2011.

[6]   Climate Data Web Site, São José dos Campos Climate, Online. pt.climate-data.org/america-do-sul/brasil/sao-paulo/sao-jose-dos-campos-6151. Accessed on: 21 Jan. 2020.

[7]   Bonotto, D.M., Comportamento hidrogeoquímico do ^{222}Rn e isótopos de urânio ^{238}U e ^{234}U sob condições controladas de laboratório e em sistemas naturais. Post-PhD thesis. São Paulo State University, Rio Claro, p. 223, 1996.

[8] HACH, *Water Analysis Handbook*, 4th ed., Hach Company: Loveland, Colorado, 2000.

[9] Hem, J.D., *Study and Interpretation of the Chemical Characteristics of Natural Water.* U.S.G.S. Water Supply Paper, 1473, pp. 1–269, 1959.

[10] Ministério da Saúde Brasileiro, Portaria N° 2914 de 12 de dezembro 2011. Procedimentos de controle e de vigilância da qualidade da água para consumo humano e seu padrão de potabilidade, p. 33, 2011.

[11] Piper, A., A graphic procedure in the geochemical interpretation of water-analyses. *Transactions of the American Geophysical Union*, **25**(6), pp. 914–928, 1944.

ABATEMENT OF MICROPLASTICS FROM MUNICIPAL EFFLUENTS BY TWO DIFFERENT WASTEWATER TREATMENT TECHNOLOGIES

JAVIER BAYO, JOAQUÍN LÓPEZ-CASTELLANOS & SONIA OLMOS
Department of Chemical and Environmental Engineering, Technical University of Cartagena, Spain

ABSTRACT

This paper discusses the role of two different wastewater treatment technologies in the removal efficiency of microplastics from the final effluent of an urban wastewater treatment plant; i.e., a membrane bioreactor as an advanced wastewater treatment technology and rapid sand filtration as a tertiary treatment after an activated sludge process. Fourteen different polymer types were identified in all wastewater samples, mainly represented by low-density polyethylene (71.89%), high-density polyethylene (5.44%), acrylate (5.24%), polypropylene (5.22%), polystyrene (4.24%), and nylon (2.56%). The main forms isolated were fibres (61.2%), followed by films (31.5%), fragments (6.7%), and beads (0.6%). The main size interval corresponded to 1 and 2 mm, accounting for 28.2%, 37.3%, and 36.8% for the influent, membrane bioreactor, and rapid sand filtration, respectively. A total of 480 microplastic particles were isolated in all wastewater samples, with an average concentration of 2.16 ± 0.42 items l^{-1}, and removal rate percentages of 79.01% for the membrane bioreactor and 75.49% for rapid sand filtration. Both technologies proved to be more efficient removing particulate microplastics (98.83% and 95.53%, respectively) than fibres (57.61% and 53.83%, respectively), showing a clear escape into the aquatic environment for fibres. The average microplastic size displayed a statistically significant increase from the influent of the sewage treatment plant (1.05 ± 0.05 mm), to rapid sand filtration effluent (1.15 ± 0.08 mm) and membrane bioreactor (1.39 ± 0.15 mm) ($F\text{-}test = 4.014$, $p = 0.019$) indicating the fibre selection made by advanced treatment technologies previously discussed.
Keywords: microlitter, microplastics, MBR, RSF, wastewater.

1 INTRODUCTION

Microplastics as a global water pollutant was identified a long time ago, since Carpenter and Smith [1] first described the presence of tiny plastic particles floating in the Sargasso Sea, a region in the North Atlantic Ocean without land boundaries, although the term "microplastic" was first introduced in scientific references by Thompson et al. [2]. They consist of plastic particles smaller than 5 mm, and are classified as primary and secondary microplastics; first of them intentionally produced in that size, and secondary ones generated from the degradation of larger plastic fragments [3]. Its incredibly mobility and ubiquity in the environment represents a global threaten, affecting, of course, the oceans, but also surface waters [4], soils [5], sediments [6], food [7], [8], drinking water [9] or air [10]. Their potential toxicological effect, acting as vectors of both organic and inorganic additives and chemicals [11], [12] are also important reasons for their removal in the environment, besides the presence of residual monomers, as styrene monomer in polystyrene microplastic particles [13].

Wastewater treatment plants (WWTPs) act as a sink or solution to remove microplastics, but also as a source because of wastewater effluents load. Several sources have been pointed out as responsible for the presence of micropalstics in WWTPs, both domestic and non-domestic ones. Domestic washing machines are prone to release fibres from synthetic textiles. Browne et al. [14] reported a production of 1,900 fibres per wash by a single garment, Almroth et al. [15] found average concentrations of 7,360 fibres m^{-2} l^{-1} in polyester fleece fabrics and 110,000 fibres per garment and wash for PET fleece, and De Falco et al. [16]

WIT Transactions on Ecology and the Environment, Vol 242, © 2020 WIT Press
www.witpress.com, ISSN 1743-3541 (on-line)
doi:10.2495/WP200021

reported an average of 6,000,000 fibres from typical 5 kg was load of polyester fabrics. Besides microplastics released from synthetic fabrics, the use of microspherules in personal care products as a substitute for natural scrubs and with a concentration ranging from 0.5 to 5% [17], may also be released as microbeads from WWTPs. Moreover, paint scraps, pellets from plastic industries, tire wreckage or microparticles from plastics consumer goods indeed reach the sewage treatment, causing the WWTPs to act both as a source and a sink for these micropollutants [18], [19].

The efficiency of different treatment technologies for the removal of microplastics in wastewater has also been reported, although some of them on a pilot scale or with short periods of analysis, that could not reveal actual seasonal variations in their counts [20]. Membrane bioreactor (MBR) has proved to be an established technology for the treatment of both municipal and industrial wastewater, with a high mixed-liqueur suspended solids concentration that may benefit nitrifying and other slow-growing microorganisms [21]. This technology has been tested for microplastic removal, both in real WWTPs [22] or in a pilot-scale bioreactor [20]. On the other hand, rapid sand filtration (RSF) as a tertiary treatment in WWTPs has also been tested. Hidayaturrahmanh and Lee [3] reported the smallest rate of microplastic removal when compared with the use of ozone as a strong oxidant or a membrane disc-filter.

Microplastics in WWTPs have been recognized only by visual identification [3], [23], or with the aid of spectrometric methods; i.e., Raman [20] or μ-Raman [24] spectroscopy, or Fourier transform infrared spectroscopy (FTIR) [19] or μ-FTIR [25], analysing all isolated microparticles [19] or just a part of them [26]. In our study, we present the removal rates of microplastics and treatment efficiencies of two different wastewater treatment technologies; i.e. MBR and RSF, both applied to the same influent in a sewage treatment plant located in the Southeast of Spain. Analyses were carried out with visual identification and FTIR confirmation.

2 MATERIALS AND METHODS

2.1 Wastewater treatment facilities

The sewage treatment plant, named "WWTP Águilas" (Fig. 1) is a full-scale plant treating both domestic and industrial wastewater. It is located in the Region of Murcia (Southeast Spain) (37°25'29''N, 01°34'46''W), with a local orography which requires the use of 19 pumping stations, distributed in two large areas and named: (1) Trasiego; (2) La Cola Calle; (3) La Cola Carretera; (4) La Cola Playa; (5) Calabardina I; (6) Calabardina II; (7) Intermedio; (8) Delicias; (9) Arqueta Entrada Paseo Delicias; (10) Hornillo Playa; (11) Hornillo Calle; (12) Renfe; (13) Rubial; (14) Las Lomas; (15) Calarreona Hotel; (16) Calarreona Playa; (17) Marqués; (18) Matadero; and (19) Calarreona Camping. In order to avoid anomalies in the operation because of a power outage, there are seven emergency power generator sets.

This plant treats approximately 12,000 m^3 d^{-1} of municipal wastewater for 29,777 equivalent inhabitants, and its treatment processes include two different lines:

(a) Conventional line – it consists of a conventional activated sludge process, with:

- Pretreatment: Initial screening with 2 manual-cleaning bars (3 cm clearance for rough solids) and 3 self-cleaning sieves (6 mm clearance for fine solids).
- Grit and grease separation with aeration.

- Primary clarification: Two circular settlers (18 m diameter). A reduction of 30% BOD_5 and 65% suspended solids is achieved. Primary sludge is pumped to a sludge gravity thickener.
- Biological reactor: Two rectangular bioreactors with a total volume of 3,132 m^3 and 6 aerators with a total power of 275 kW.
- Secondary clarification: Two circular gravity clarifiers (22 m diameter). Settled sludge is collected and partially pumped (70%) to the biological reactor. Secondary sludge is sent to the gravity thickener and clarified wastewater to an Accelator® settler (14 m diameter) prior to rapid sand filtration.
- Tertiary treatment: Three parallel open rapid sand filters with two units per filter and a central channel. Each filter is 6.04 m in length and 5.86 m in width, including the central channel. The filtration speed is 8 m h^{-1} and the maximum loss due to filter washing is 1.5%. the total filtration surface is 75 m^2 (Fig. 2).

(b) Membrane bioreactor line: It consists of an advanced wastewater treatment with microfiltration membranes:

- Pretreatment: Initial screening with 2 self-cleaning bars (2 cm clearance for rough solids) and 2 self-cleaning sieves (3 mm clearance for fine solids). Sands are gravity removed.
- Lamination tank: It allows a 24-hours lamination flow with a total volume of 1,800 m^3, pumping wastewater into an anoxic chamber.
- Anoxic chamber: One anoxic chamber with a volume of 363,5 m^3, where denitrification is partially carried out.
- Biological reactor: One rectangular bioreactor with a volume of 1,050 m^3 with three parts: aerobic, anaerobic, and facultative zones.
- Membrane tank: This unit has a total volume of 315 m^3 with 10 submerged flat-sheet membrane modules EK-400 (Kubota Corporation, Japan) comprising a total of 4,000 membranes distributed into two lines at different heights within the membrane tank, and 3,560 m^2 of effective microfiltration surface. The designed flow is 1,800 m^3 d^{-1} and permeated wastewater is directly transferred for agricultural use.

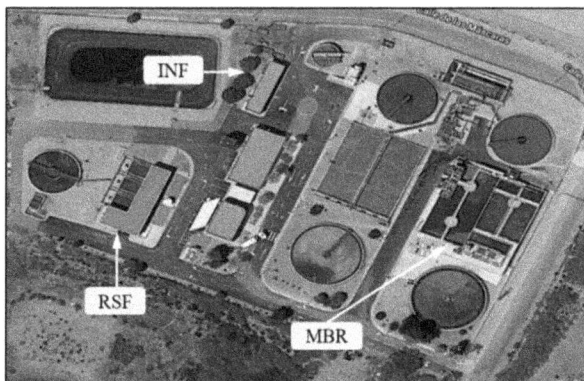

Figure 1:   Operation of both wastewater treatment lines in WWTP Águilas, indicating where samples are collected: (INF) Influent; (MBR) Membrane bioreactor; (RSF) Rapid sand filter.

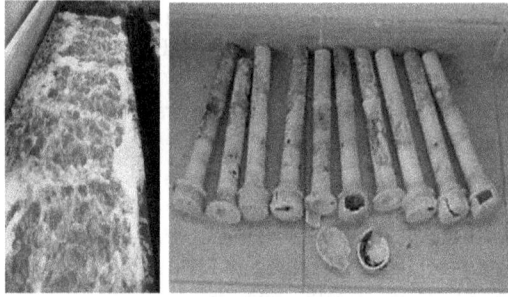

Figure 2: Rapid sand filtration.

Table 1: Sampling days and volumes.

Sampling day	INF (l)	MBR (l)	RSF (l)
14 February 2018	3.78	3.50	3.56
14 March 2018	4.98	-	3.75
11 April 2018	3.20	5.49	5.24
10 May 2018	3.97	4.93	5.35
7 June 2018	4.41	4.83	4.47
5 July 2018	4.62	4.71	4.80
1 August 2018	3.42	3.77	3.87
13 September 2018	3.82	3.78	3.89
11 October 2018	4.25	3.82	3.86
8 November 2018	5.37	3.81	3.84
13 December 2018	3.72	3.40	3.41
17 January 2019	3.32	3.86	3.87
14 February 2019	5.20	4.71	4.77
26 March 2019	4.23	4.11	3.98
10 April 2019	3.91	3.41	3.33
16 May 2019	3.81	4.12	3.74
12 June 2019	3.78	3.84	3.34
18 July 2019	3.68	9.20	3.77
Average	4.18	4.32	4.04
Total	73.47	75.29	72.74

2.2 Sample collection and processing

The sewage plant has been monitored for 18 months with a total of 53 grab samples (Table 1); i.e., from 14th February 2018 to 18th July 2019, and accounting for 73.47 l from INF, 75.29 l from MBR, and 72.74 l from RSF.

Wastewater samples were always collected in the morning, between 9:00 h and 11:00 h, in glass containers with metallic lid. Samples from MBR and RSF were directly vacuum filtered trough a Büchner funnel by means of a membrane filter (Prat Dumas, Couze-St-Front, France, 110 mm Ø pore size 0.45 μm), and microplastics included in INF samples were previously isolated by density separation with a 120 g l^{-1} NaCl solution (density 1.08 g ml^{-1}) as previously reported in Bayo et al. [19]. The mixture was placed into a 2 l glass beaker with mechanical stirring for 20 min. Supernatant with floating particles was filtered through the same membrane filter and, after it was washed with bi-distilled water, content in the Petri dish was dried overnight in an air-forced stove.

In order to mitigate the risk of pollution, nitrile gloves and clean 100% cotton lab gowns were worn by analysts. Glass Petri dishes, both 40 mm Ø and 120 mm Ø, were used and lab plastic devices were limited to the maximum, replacing plastic caps with aluminum foil when necessary. Laboratory benches and glassware were always twice washed with bi-distilled before each experiment. In order not to act as a source of microplastics, polyethylene containers for bi-distilled water were examined twice during the whole sampling campaign by vacuum filtering 1.5 l of stored content. No microplastics were isolated from blank samples.

Possible microplastic particles were examined under an Olympus SZ-61TR Zoom Trinocular Microscope (Olympus Co., Tokyo, Japan), providing a superior image quality with a 10° convergence angle at a working distance of 110 mm, magnification range from 6.7x to 45x and LED lighting. This trinocular microscope was coupled to a Leica MC190 HD digital camera, with a maximum resolution of 1596 x 1196 pixels, 10 bits per color channel, 7.5 frames per second at full resolution, and 0.1 ms to 1 s exposure time.

The infrared spectra were acquired with a Thermo Nicolet 5700 Fourier transformed infrared (FTIR) spectrometer (Thermo Nicolet Analytical Instruments, Madison, WI, USA), provided with a deuterated triglycine sulfate (DTGS) detector and KBr detector. The spectra were collected with an average of 20 scans and a resolution of 16 cm^{-1} in the range of 4000–400 cm^{-1} wavelength. Spectra were controlled and evaluated by the OMNIC software package, by means of a reference polymer library containing spectra of all common polymers, together with literature [27]. Data were processed with the SPSS (Statistic Package for Social Science) 26.0 software.

3 RESULTS AND DISCUSSION

3.1 Polymer types identified in wastewater samples

Different types of microplastics were isolated in the INF, MBR, and RSF, and some microscopic images are depicted in Figs 3, 4, and 5, respectively. When microplastics from all sampling points were considered, a total of 14 different polymers were identified, all of them in the influent except for melamine (MUF) that was only identified in a MBR sample. The percentage of each polymer type found in the INF was: low-density polyethylene (LDPE) (71.89%), followed by high-density polyethylene (HDPE) (5.44%), acrylate (AC) (5.24%), polypropylene (PP) (5.22%), polystyrene (PS) (4.24%), polyamide or nylon (NYL) (2.56%), methacrylate (MCR) (1.76%), poly(ethylene:propylene) (EPM) (1.07%), biopolymer (BPL) (0.62%), polyester (PEST) (0.56%), polyvinyl (PV) (0.50%), polyisobutylene (PIB) (0.44%), and Teflon (PTFE) (0.44%). Lares et al. [20] reported similar results; i.e., 63.9% of polyethylene in microplastic particles, from a municipal WWTP located in Finland with activated sludge process and a pilot-scale membrane bioreactor, and Long et al. [28] reported PP (30.2%), PE (26.9%), and PS (10.3%) in the influent of seven

WWTPs in China. According to Plastics Europe [29], the demand by resin type during 2018 was leaded by the polyolefins polyethylene and polypropylene, and these, plus polyester, are the major components of microplastics found in the aquatic environment [30].

In our study, the presence of acrylate in INF; i.e., poly(lauryl acrylate), poly(cyclohexyl acrylate), and poly(11-bromoundecyl acrylate), indicates its wide use in daily products, as commercial shower gels, peelings, waterproof sunscreen, or as a gelling agent in lipsticks and paint particles [31], [32]. As previously indicated, MUF was only isolated in a MBR sample, and RSF displayed three types of polymer: LDPE (58.67%), PV (27.55%), and NYL (13.78%). Analysis of variance confirmed that the removal of LDPE from INF (1.69 ± 0.59 items l^{-1}) to RSF (0.06 ± 0.04 items l^{-1}) was statistically significant ($F\text{-}test$ = 7.634, p = 0.001). Other polymer types, as EPM and PV, previously reported in relatively high number in beach sediments [33] because their use in packaging and shipbuilding, respectively, only represented 1.07% and 0.50% in wastewater samples.

Some studies have pointed out that the enormous variability of polymeric plastic microparticles reported in different WWTPs has also to do with the use of oxidation processes reported to digest the organic matter in wastewater samples. In this sense, Carr et al. [34] found a potential loss of PE and PP after a digestion process, and Munno et al. [35] reported the loss of polyamide after a wet oxidation process with temperature higher than 60°C.

The presence of styrene-butadiene rubber (SBR), the main component of car tires, was hardly detected in microplastics (0.009 items l^{-1}), despite being suggested as a potential marker from car tire abrasion, the main source of microplastics into the environment [36]. Microparticles from car tire wear have a wide range of densities, as reported by Leads and Weinstein [37], depending on whether they come from raw recycled tire crumb rubber (1.13–1.16 g ml^{-1}) [38] or aggregated with road dust and mineral particles (up to 1.8 g ml^{-1}) [39], [40]. SBR is also difficult to be quantified by FTIR, due to the carbon black added as a filler [41], and rainfall scarcity in our Region could also contribute to its low detection, as the major fraction of road dust-associated microplastic particles is expected to be found in the runoff from the road verge generated during rainfall events [42].

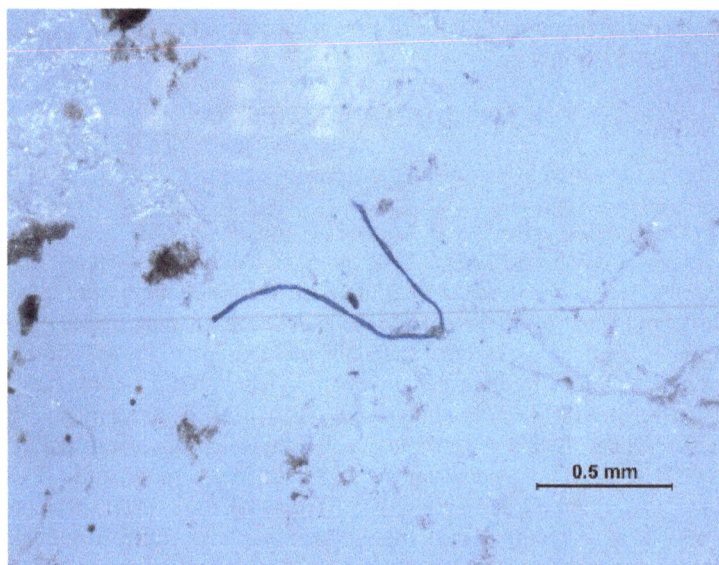

Figure 3: Nylon (NYL) blue fibre from influent (INF) (0.86 mm, 13 September 2018).

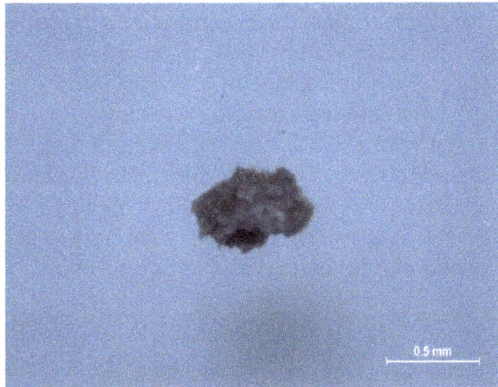

Figure 4:   Melamine (MUF) brown fragment from membrane bioreactor (MBR) (0.68 mm, 7 June 2018).

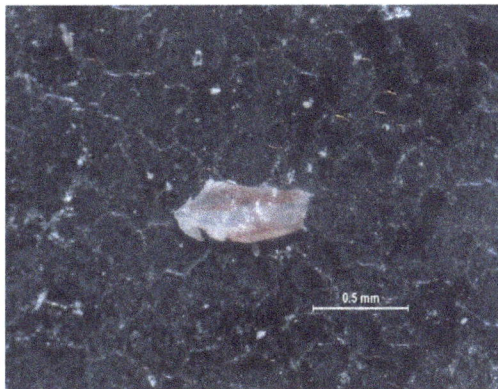

Figure 5:   Low-density polyethylene (LDPE) pink film from rapid sand filter (RSF) (0.70 mm, 13 September 2018).

3.2 Shapes, sizes and colours of isolated microplastics

The use of the stereomicroscope allowed a more exhaustive classification according to the shape of microplastics, with fibres (MFBs) (61.2%) being the most recurrent form of microplastics in all wastewater samples (Fig. 3), followed by films (31.5%) (Fig. 5), fragments (6.7%) (Fig. 4), and beads (0.6%) (Fig. 6), which represents a 38.8% of microplastic forms (MPPs) different to MFBs. These results are similar to that reported by Lares et al. [20] in a pilot-scale bioreactor operated in a WWTP in Finland. We also observed an increase in the ratios MFBs/Total microplastics and MPPs/Total microplastics from INF, accounting for 48.09% and 51.91%, respectively, to MBR, 96.72% and 3.28%, and RSF, 90.79% and 9.21%, respectively. Because of the absence of plastic industries nearby the studied sewage treatment plant, it is clear that clothing fibres from washing machine effluents can bypass the treatment processes and escape into the aquatic environment [43], and this release has to do with many other factors; i.e., textile properties, washing conditions, or type of detergent and softener used [10], [14]–[16].

A 70.2% of microplastics larger than 1 mm corresponded to MFBs., and 58.9% of total microplastics were under 1 mm. Only after MBR, microplastics smaller than 200 µm were isolated. As depicted in Fig. 7, the percentage of each size interval was different according to the considered sampling point into the wastewater treatment plant. The main size range in all considered stages was between 1 and 2 mm, accounting for a 28.2%, 37.3%, and 36.8% for INF, MBR, and RSF, respectively. Average microplastic size displayed a statistically significant increase from INF (1.05 ± 0.05 mm), to RSF (1.15 ± 0.08 mm) and MBR (1.39 ± 0.15 mm) (*F-test* = 4.014, *p* = 0.019) indicating the fibre selection made by advanced treatment technologies previously discussed.

Figure 6: Polystyrene (PS) white bead from influent (INF) (1.10 mm, 14 February 2018).

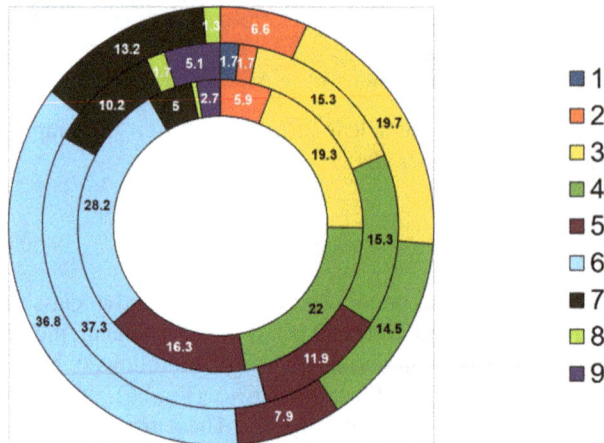

Figure 7: Size ranges according to the Spanish Environmental Ministry: (1) 0.2 mm; (2) 0.2–0.4 mm; (3) 0.4–0.6 mm; (4) 0.6–0.8 mm; (5) 0.8–1.0 mm; (6) 1.0–2.0 mm; (7) 2.0–3.0 mm; (8) 3.0–4.0 mm; and (9) 4.0–5.0 mm. Inner ring means RSF; Medium ring means MBR; Outer ring means INF (results expressed as a percentage).

Most fragments were categorized as opaque (71.9%), and most films as transparent (70.9%). These results are similar to that reported by Leslie et al. [44] in WWTPs from The Netherlands. In the case of film forms, the average concentration decreased from INF (1.7 ± 0.6 items l⁻¹) to RSF (0.1 ± 0.0 items l⁻¹), totally disappearing in MBR effluent (*F-test* = 6.596, *p* = 0.003).

3.3 The removal rates of MBR and RSF

A total of 480 microplastics were isolated from all wastewater samples, with an average concentration of 2.16 ± 0.42 items l⁻¹. They represented 76.68% of total isolated microparticles, which corroborates the need of specific spectrometry techniques; i.e., FTIR, to successfully identify microplastics out of all types of microparticles captured in these studies [19], including calcium stearate, glycerine and lipid mediators from solidified soap, silicates, chipboard fragments, and so on [45]. Nevertheless, matching with IR spectra of standardized polymers from commercially available selected spectral libraries could be difficult, due to weathered and polluted surfaces of microplastics and interference with other compounds present in wastewater samples [46].

Statistically significant differences were observed between the average concentration of microplastics collected in INF (4.40 ± 1.01 items l⁻¹) versus micropalstics collected in MBR (0.92 ± 0.21 items l⁻¹) and RSF (1.08 ± 0.28 items l⁻¹) (*F-test* = 9.953, *p* = 0.000), indicating a clear removal with both technologies. The removal percentage for MBR was 79.01%, higher than for RSF (75.49%), although there were no statistically significant differences between both technologies (*F-test* = 0.195, *p* = 0.661). Although both MPPs and MFBs proved to decrease through the sewage treatment plant, they displayed different removal rates. MPPs showed a statistically significant decrease from INF (2.26 ± 0.70 items l⁻¹) to MBR (0.03 ± 0.02 items l⁻¹) and RSF (0.10 ± 0.05 items l⁻¹) (*F-test* = 9.454, *p* = 0.000), accounting for a 98.83% and 95.53% removal, respectively. In contrast, changes in MFBs through the WWTP were smaller, from INF (2.12 ± 0.55 items l⁻¹) to MBR (0.90 ± 0.21 items l⁻¹) (57.61%) and RSF (0.98 ± 0.27 items l⁻¹) (53.83%), still with statistically significant differences (*F-test* = 3.214, *p* = 0.049). Therefore, although a good performance in MPPs removal is achieved by both technologies, MFBs still bypass and escape into the aquatic environment, resulting in a global removal rate lower than that reported for a conventional activated sludge process [19]. Some reasons have been proposed to explain this fact; Leslie et al. [44] indicated that the high pressure applied to a MBR system could favor this escape, as well as their small size and morphology could enable them to longitudinally pass through the RSF [22], [45]. Also, a direct air pollution with MFBs derived from apparel articles and household dust released in open-air sewage treatment plants could be an important fact [47].

4 CONCLUSIONS

The results of this study allow us to conclude that advanced technologies, as membrane bioreactor, and tertiary treatments, as rapid sand filtration, proved to be more efficient removing particulate microplastics than microfibres, achieving a global microplastic elimination of 79.01% and 75.49%, respectively, lower than that reported for conventional activated sludge processes; i.e., 90.3% [19], and without statistical significant differences between them. As much as 14 different polymer types were isolated in wastewater samples, indicating the large amount of urban and industrial plastic sources used today, reaching the wastewater treatment plants in the form of microplastic particles, and matching to the demand by resin type indicated by the leading pan-European plastic trade association.

ACKNOWLEDGEMENT
Authors sincerely acknowledge the grant from "Fundación Séneca" (20268/FPI/17) for the financial support of predoctoral student Sonia Olmos.

REFERENCES
[1]   Carpenter, E.J. & Smith, K.L., Plastics on the Sargasso Sea surface. *Science*, **175**(4027), pp. 1240–1241, 1972.
[2]   Thompson, R.C. et al., Lost at sea: Where is all the plastic? *Science*, **304**(5672), pp. 838–838, 2004.
[3]   Hidayaturrahman, H. & Lee, T.G., A study on characteristics of microplastic in wastewater of South Korea: Identification, quantification, and fate of microplastics during treatment process. *Marine Pollution Bulletin*, **146**, pp. 696–702, 2019.
[4]   Eriksen, M. et al., Microplastic pollution in the surface waters of the Laurentian Great Lakes. *Marine Pollution Bulletin*, **77**(1–2), pp. 177–182, 2013.
[5]   He, D., Luo, Y., Lu, S., Liu, M., Song, Y. & Lei, L., Microplastics in soils: analytical methods, pollution characteristics and ecological risks. *TrAC Trends in Analytical Chemistry*, **109**, pp. 163–172, 2018.
[6]   Peng, G., Zhu, B., Yang, D., Su, L., Shi, H. & Li, D., Microplastics in sediments of the Changjiang Estuary, China. *Environmental Pollution*, **225**, pp. 283–290, 2017.
[7]   Iñiguez, M.E., Conesa, J.A. & Fullana, A., Microplastics in Spanish table salt. *Scientific Reports*, **7**(1), pp. 1–7, 2017.
[8]   Toussaint, B. et al., Review of micro-and nanoplastic contamination in the food chain. *Food Additives & Contaminants: Part A*, **36**(5), pp. 639–673, 2019.
[9]   Mintenig, S.M., Löder, M.G.J., Primpke, S. & Gerdts, G., Low numbers of microplastics detected in drinking water from ground water sources. *Science of the Total Environment*, **648**, pp. 631–635, 2019.
[10]   Prata, J.C., Airborne microplastics: consequences to human health? *Environmental Pollution*, **234**, pp. 115–126, 2018.
[11]   Bayo, J., Guillén, M., Olmos, S., Jiménez, P., Sánchez, E. & Roca, M.J., Microplastics as vector for persistent organic pollutants in urban effluents: The role of polychlorinated biphenyls. *International Journal of Sustainable Development and Planning*, **13**(4), pp. 671–682, 2018.
[12]   Bayo, J., Olmos, S. & López-Castellanos, J., Non-polymeric chemicals or additives associated with microplastic particulate fraction in a treated urban effluent. *WIT Transactions on The Built Environment*, **179**, pp. 303–314, 2018.
[13]   Andrady, A.L., *Environmental Sustainability of Plastics*, John Wiley, 2016.
[14]   Browne, M.A. et al., Accumulation of microplastic on shorelines worldwide: sources and sinks. *Environmental Science & Technology*, **45**(21), pp. 9175–9179, 2011.
[15]   Almroth, B.M.C., Åström, L., Roslund, S., Petersson, H., Johansson, M. & Persson, N. K., Quantifying shedding of synthetic fibers from textiles; a source of microplastics released into the environment. *Environmental Science and Pollution Research*, **25**(2), pp. 1191–1199, 2018.
[16]   De Falco, F. et al., Evaluation of microplastic release caused by textile washing processes of synthetic fabrics. *Environmental Pollution*, **236**, pp. 916–925, 2018.
[17]   Bayo, J., Martínez, A., Guillén, M., Olmos, S., Roca, M.J. & Alcolea, A., Microbeads in commercial facial cleansers: threatening the environment. *Clean–Soil, Air, Water*, **45**(7), pp. 1–11, 2017.

[18] Bayo, J., Olmos, S., López-Castellanos, J. & Alcolea, A., Microplastics and microfibers in the sludge of a municipal wastewater treatment plant. *International Journal of Sustainable Development and Planning*, **11**(5), pp. 812–821, 2016.

[19] Bayo, J., Olmos, S. & López-Castellanos, J., Microplastics in an urban wastewater treatment plant: The influence of physicochemical parameters and environmental factors. *Chemosphere*, **238**, 124593, 2020.

[20] Lares, M., Ncibi, M.C., Sillanpää, M. & Sillanpää, M., Occurrence, identification and removal of microplastic particles and fibers in conventional activated sludge process and advanced MBR technology. *Water Research*, **133**, pp. 236–246, 2018.

[21] Dvořák, L., Svojitka, J., Wanner, J. & Wintgens, T., Nitrification performance in a membrane bioreactor treating industrial wastewater. *Water Research*, **47**(13), pp. 4412–4421, 2013.

[22] Talvitie, J., Mikola, A., Koistinen, A. & Setälä, O., Solutions to microplastic pollution– Removal of microplastics from wastewater effluent with advanced wastewater treatment technologies. *Water Research*, **123**, pp. 401–407, 2017.

[23] Michielssen, M.R., Michielssen, E.R., Ni, J. & Duhaime, M.B., Fate of microplastics and other small anthropogenic litter (SAL) in wastewater treatment plants depends on unit processes employed. *Environmental Science: Water Research & Technology*, **2**(6), pp. 1064–1073, 2016.

[24] Gündoğdu, S., Çevik, C., Güzel, E. & Kilercioğlu, S., Microplastics in municipal wastewater treatment plants in Turkey: a comparison of the influent and secondary effluent concentrations. *Environmental Monitoring and Assessment*, **190**(11), 626 pp., 2018.

[25] Tagg, A.S., Sapp, M., Harrison, J.P. & Ojeda, J.J., Identification and quantification of microplastics in wastewater using focal plane array-based reflectance micro-FT-IR imaging. *Analytical Chemistry*, **87**, pp. 6032–6040, 2015.

[26] Edo, C., González-Pleiter, M., Leganés, F., Fernández-Piñas, F. & Rosal, R., Fate of microplastics in wastewater treatment plants and their environmental dispersion with effluent and sludge. *Environmental Pollution*, **259**, pp. 113837, 2020.

[27] Hummel, D.O., *Atlas of Plastics Additives: Analysis by Spectrometric Methods*, Springer: Berlin/Heidelberg, 537 pp., 2002.

[28] Long, Z. et al., Microplastic abundance, characteristics, and removal in wastewater treatment plants in a coastal city of China. *Water Research*, **155**, pp. 255–265, 2019.

[29] Plastics – the Facts 2018, An analysis of European plastics production, demand and waste data. www.plasticseurope.org/application/files/6315/4510/9658/Plastics_the_facts_2018_AF_web.pdf. Accessed on: 3 Feb. 2020.

[30] Lee, H., Shim, W.J. & Kwon, J.H., Sorption capacity of plastic debris for hydrophobic organic chemicals. *Science of the Total Environment*, **470**, pp. 1545–1552, 2014.

[31] Liebezeit, G. & Dubaish, F., Microplastics in beaches of the East Frisian islands Spiekeroog and Kachelotplate. *Bulletin of Environmental Contamination and Toxicology*, **89**(1), pp. 213–217, 2012.

[32] Chae, D.H., Kim, I.S., Kim, S.K., Song, Y.K. & Shim, W.J., Abundance and distribution characteristics of microplastics in surface seawaters of the Incheon/Kyeonggi coastal region. *Archives of Environmental Contamination and Toxicology*, **69**(3), pp. 269–278, 2015.

[33] Bayo, J., Rojo. D. & Olmos, S., Abundance, morphology and chemical composition of microplastics in sand and sediments from a protected coastal area: The Mar Menor lagoon (SE Spain). *Environmental Pollution*, **252**, pp. 1357–1366, 2019.

[34] Carr, S.A., Liu, J. & Tesoro, A.G., Transport and fate of microplastic particles in wastewater treatment plants. *Water Research*, **91**, pp. 174–182, 2016.

[35] Munno, K., Helm, P.A., Jackson, D.A., Rochman, C. & Sims, A., Impacts of temperature and selected chemical digestion methods on microplastic particles. *Environmental Toxicology and Chemistry*, **37**(1), pp. 91–98, 2018.

[36] Sommer, F. et al., Tire abrasion as a major source of microplastics in the environment. *Aerosol and Air Quality Research*, **18**, pp. 2014–2028, 2018.

[37] Leads, R.R. & Weinstein, J.E., Occurrence of tire wear particles and other microplastics within the tributaries of the Charleston Harbor Estuary, South Carolina, USA. *Marine Pollution Bulletin*, **145**, pp. 569–582, 2019.

[38] Rhodes, E.P., Ren, Z. & Mays, D.C., Zinc leaching from tire crumb rubber. *Environmental Science & Technology*, **46**(23), pp. 12856–12863, 2012.

[39] Klöchner, P., Reemtsma, T., Eisentraut, P., Braun, U., Ruhl, A.S. & Wagner, S., Tire and road wear particles in road environment–quantification and assessment of particle dynamics by Zn determination after density separation. *Chemosphere*, **222**, pp. 714–721, 2019.

[40] Unice, K.M., Kreider, M.L. & Panko, J.M., Comparison of tire and road wear particle concentrations in sediment for watersheds in France, Japan, and the United States by quantitative pyrolysis GC/MS analysis. *Environmental Science & Technology*, **47**(15), pp. 8138–8147, 2013.

[41] Haave, M., Lorenz, C., Primpke, S. & Gerdts, G., Different stories told by small and large microplastics in sediment-first report of microplastic concentration in an urban recipient in Norway. *Marine Pollution Bulletin*, **141**, pp. 501–513, 2019.

[42] Vogelsang, C. et al., Microplastics in road dust-characteristics, pathways and measures. *Report from Norwegian Institute for Water Research*, pp. 170, 2019.

[43] Ladewig, S.M., Bao, S. & Chow, A.T., Natural fibers: A missing link to chemical pollution dispersion in aquatic environments. *Environmental Science & Technology*, **49**, pp. 12609–12610, 2015.

[44] Leslie, H.A., Brandsma, S.H., Van Velzen, M.J.M. & Vethaak, A.D., Microplastics en route: Field measurements in the Dutch river delta and Amsterdam canals, wastewater treatment plants, North Sea sediments and biota. *Environmental International*, **101**, pp. 133–142, 2017.

[45] Ziajahromi, S., Neale, P.A., Rintoul, L. & Leusch, F.D., Wastewater treatment plants as a pathway for microplastics: Development of a new approach to sample wastewater-based microplastics. *Water Research*, **112**, pp. 93–99, 2017.

[46] Gies, E.A. et al., Retention of microplastics in a major secondary wastewater treatment plant in Vancouver, Canada. *Marine Pollution Bulletin*, **133**, pp. 553–561, 2018.

[47] Zobkov, M. & Esiukova, E., Microplastics in Baltic bottom sediments: Quantification procedures and first results. *Marine Pollution Bulletin*, **114**(2), 724–732, 2017.

MICROBIOME ANALYSIS OF THE BACTERIAL POPULATION IN A BENCH-SCALE-ACTIVATED SLUDGE REACTOR EXPOSED TO AN ARTIFICIAL INSECTICIDE SPILL

ÁNGELA BAEZA-SERRANO[1], MARIA JOSÉ TÁRREGA[2], JUAN F. MARTÍNEZ-BLANCH[3],
ANTONIA ROJAS[4], MARTA TORTAJADA[4], GLORIA FAYOS[5] & TATIANA MONTOYA[1]
[1]Global Omnium Medio Ambiente, Spain
[2]Empresa General Valenciana del Agua, Spain
[3]ADM-Lifesequencing – Health & Wellness – ADM Nutrition, Spain
[4]ADM-Biopolis – Health & Wellness – ADM Nutrition, Spain
[5]Aguas de Valencia, Spain

ABSTRACT

Wastewater treatment plants (WWTPs) must process wastewater efficient and continuously. Uncontrolled toxic spills can damage the activated sludge units, reducing the treatment capacity of the bacterial culture. Consequently, water below the quality requirements could be released to the receiving environment, contributing to the loss of biological diversity, degradation of water resources and generating public health threats. The objective of this assay was to study the changes that an insecticide spill produces in the activated sludge microbiome of a laboratory-scale reactor. Next-generation sequencing was carried out to idntify species that could serve as key indicators of a negative biological process affection. The bench-scale system consisted of a vessel with primary settled wastewater feeding a 10-L biological reactor with intermittent aeration cycles to remove organic matter and nutrients. A secondary clarifier with manual purging and an external recirculation to the biological reactor completed the system. Both settled wastewater and activated sludge was obtained from the same WWTP. An insecticide containing D-tetramethrin, cyphenothrin and pyriproxyfen was previously proven to decrease significantly the nitrification activity of activated sludge with a 2 mL dose. The experiment was repeated with biological sampling to monitor activated sludge microbiome 2 hours after the spill, and on days 1, 2, 3, 6 and 7. The insecticide's toxic effect on the biological process was demonstrated to be significant, changing the distribution of the activated sludge population. Results show that relative abundance of certain groups increased 2 hours after the spill, with one species specifically being very sensitive to the insecticide effect, disappearing completely after only 2 hours. These results show that some microorganism groups could be monitored as indicators of a negative affection in the activated sludge biological process and have a potential use to develop early detection kits for biological failures in WWTPs.
Keywords: activated sludge, metagenomic analysis, toxic spills, bacterial community.

1 INTRODUCTION

The activated sludge is the most extended secondary wastewater treatment [1]. It is a biological process where a bacterial culture grows in a biological reactor mostly under aerobic conditions. This culture is able to metabolize organic compounds, nutrients (nitrogen and phosphorus) and other substances present in wastewater.

Knowledge of the microorganisms involved in the purification processes in activated sludge is crucial to the development of operation strategies that ensure the right treatment of sewage and, therefore, the environment protection.

Classical methods for microorganisms detection of environmental samples, like microscopy or culture-depending methods, have been replaced by metagenomic approaches, like next-generation sequencing. High throughput sequencing targeting conserved regions in

WIT Transactions on Ecology and the Environment, Vol 242, © 2020 WIT Press
www.witpress.com, ISSN 1743-3541 (on-line)
doi:10.2495/WP200031

microbial genomes is now regarded as most reliable and cost-effective method for taxonomical identification and species composition analysis of environmental samples [2].

Wastewater treatment plants (WWTP) sometimes suffer uncontrolled toxic spills that can damage the biomass from biological systems, reducing the treatment capacity of the bacterial culture, especially regarding to its nitrifying capacity. As a result, water below the quality requirements could be released to the receiving environment, contributing to the eutrophication processes and degradation of water resources. Thus, the increase of the urban wastewater toxic load hinders the treatment. During the year, the entry of spills increases the plant load and that cannot be eliminated by the bacterial culture. These discharges are often seasonal and industrials and usually coincide with campaigns such as the pouring of both, viticulture or cannery.

The welfare state of developed countries and the associated consumerism has doubled the organic material and solids in wastewaters, and has produced the appearance of emergent water contaminants, such as heavy metals and organic molecules, some of them classified as priority substances. Moreover, many common human activities related to hygiene and health standards end up contaminating the water, for instance: Rat poison treatments in sewage systems, Leaching of insecticides (cockroaches, mosquitoes, etc.) in parks and gardens, extensive use of cosmetics, personal hygiene products and medicines in the household as well as in hospitals, nursing homes etc. In the majority of cases, when the WWTP operator detects that the organic or nutrient removal performance is decreasing, most of the biomass has already been inevitably damaged and the usual procedure to recover normal activity consists on incrementing the air supply to the bioreactor, with the consequent increase in energy consumption and economic cost overrun. Aeration systems are characterized by high energy consumption that, depending on the type and size of the process, may constitute about 50–70% of total energy consumption of a plant.

The recovery of the biological process until the normal operating conditions are re-established may take as long as days, weeks or even months. Depending on the size of the WWTP to achieve such recovery, different strategies should be used, for instance: to increase aeration, to use coagulants, to start up homogenization tanks or to reseed process, among other.

Next-generation sequencing data on microbial diversity can be used to enhance bioaugmentation and boost wastewater treatment, or to improve the biodegradation of specific contaminants [3].

In this context, the LIFE16 ENV/ES/000390 BACTIWATER project [4] proposed the use of microbial growth enhancers to recover the biological units after the entrance of uncontrolled spills to WWTP, thus boosting its ecosystem recovery. Moreover, within the project the implementation of an early detection system based in "omic" techniques to early detect malfunctions in the process by alteration of the microbial populations will be carrying out.

In this work, a first approximation of the spill effects in activated sludge population at bench-scale is presented.

2 MATERIALS AND METHODS

2.1 Pilot scale assay

The bench-scale system (Fig.1) consisted of an agitated vessel with primary settled wastewater feeding a 10-L biological reactor with intermittent aeration cycles to remove organic matter and nutrients. A secondary clarifier with manual purging and an external recirculation to the biological reactor completed the system (Table 1). Both primary settled wastewater and activated sludge were obtained from a WWTP with an activated sludge with nutrient removal system (Valencia, Spain).

To evaluate the effect of an artificial spill on the biomass into the biological reactor, a bench-scale assay was carried out. A commercial insecticide containing D-tetramethrin, cyphenothrin and pyriproxyfen was used. This insecticide was chosen because of is widely used and the toxicity of its components to various organisms, including humans, has been proven.

Biological samples were collected in order to monitor activated sludge microbiome 2 hours after the spill, and on days 1, 2, 3, 6 and 7.

Figure 1: Bench-scale system.

Table 1: Parameters of laboratory scaled plant used during the assay.

Parameter	Value	Unit
Inlet flow	0.6	l/h
SS demo plant	1.356	mg/l
BOD$_5$	110	mg/l
	0.002	kg/d
TSS	0.014	Kg
Reactor volume	0.01	m^3
F/M ratio	0.12	Kg BOD$_5$/Kg MLSS/d

Respirometry tests were carried out using BMT+ (SURCIS) respirometer at Global Omnium laboratory to evaluate the nitrification activity before and after the spill.

Mixed liquor samples were collected in 5ml Eppendorf tubes, frozen at -20°C and transported in cold boxes to Lifesequencing S.L.-ADM laboratory (Paterna, Valencia, Spain) for later metagenomic analysis.

2.2 Microbiome analysis

Genomic DNA was extracted from 2 mL of the activated sludge samples with Qiamp Power Fecal Mini kit (Qiagen) with enzymatic lysis and mechanic disruption. DNA were amplified following the 16S Metagenomic Sequencing Library Illumina 15044223 B protocol (Illumina). In brief, the first amplification step, primers were designed containing: a universal linker sequence allowing amplicons for incorporation indexes and sequencing primers by Nextera XT Index kit (Illumina); and 16S rRNA gene universal primers [5] and in the second and last amplification indexes were included. Libraries were quantified by fluorimetry using Quant-iT™ PicoGreen™ dsDNA Assay Kit (Thermofisher) and pooled before to sequencing on the MiSeq platform (Illumina), configuration 300 cycles paired reads. The size and quantity of the pool were assessed on the Bioanalyzer 2100 (Agilent) and with the Library Quantification Kit for Illumina (Kapa Biosciences), respectively. PhiX Control library (v3) (Illumina) was combined with the amplicon library (expected at 20%). Sequencing data were available within approximately 56 hours. Image analysis, base calling and data quality assessment were performed on the MiSeq instrument.

For massive sequencing, the hypervariable region V3-V4 of the bacterial 16s rRNA gene was amplified using key-tagged eubacterial primers [6] and sequenced with a MiSeq Illumina Platform, following the Illumina recommendations.

The resulting sequences were split taking into account the barcode introduced during the PCR reaction, while R1 and R2 reads were overlapped using PEAR program version 0.9.1 [7] providing a single FASTQ file for each of the samples. Quality control of the sequences was performed in different steps, (i) quality filtering (with a minimum threshold of Q20) was performed using fastx tool kit version 0.013, (ii) primer (16s rRNA primers) trimming and length selection (reads over 300 nts) was done with cutadapt version 1.4.1 [8]. These FASTQ files were converted to FASTA files and UCHIME program version 7.0.1001 was used in order to remove chimeras that could arise during the amplification and sequencing step. Those clean FASTA files were BLAST [9] against NCBI 16s rRNA database using blastn version 2.2.29+. The resulting XML files were processed using a python script developed by ADM-Lifesequencing S.L. (Paterna, Valencia, Spain) in order to annotate each sequence at different phylogenetic levels (Phylum, Family, Genus and Species).

3 RESULTS AND DISCUSSION

3.1 Bacterial community structure

Fig. 2 shows the core phylum abundance in activated sludge assay samples (taxa represented occurred at >1% abundance in at least one sample). Between 4.4–5.1% of the sequencing reads could not be assigned to any taxa at phylum level (No hit in Fig. 2).

The predominant phyla at the beginning of the assay were Proteobacteria, Actinobacteria and Bacteroidetes, corresponding to 51% of the total sequences. The same phyla were found by Meerbergen et al. [9] and Liang et al. [10], being Proteobacteria predominant in domestic sewage. However, at the end of the assay after 7 days of the biomass exposition to the insecticide, Firmicutes phylum replaces Actinobacteria as the third phylum in relative abundance. Firmicutes have been described as widely distributed in anaerobic sludge treatment systems [11] and they are versatile in degrading a big array of environmental substrates [12]. Also, Bacteroidetes, which are often reported as proteolytic bacteria, involved in degrading proteins [13], grows its abundance.

Fig. 3 shows the core genera abundance in activated sludge assay samples (taxa represented occurred at >1% abundance in at least one sample). Between 4 a 5% of the sequencing reads could not be assigned to any taxa at genus level (No hit in Fig.1), which is a significant fall comparing with the 30% described in Wang et al. [14] or the 32–34% reported by Zhang et al. [15].

The calculated species richness Chao 1 and Shannon index are shown in Table 2. Activated sludge samples present high Shannon Index value, indicating that ecosystems are very diverse. Chao 1 and Shannon index values are greater than those found by Gonzalez-Martínez et al. [16] in 10 different wastewater treatment systems in Spain and Netherlands (Chao 1 1395,003-441,150 and Shannon index 5.137-2.831). Two hours after the spill a decline of the richness is observed that could be attributed to the toxic effect of the spill.

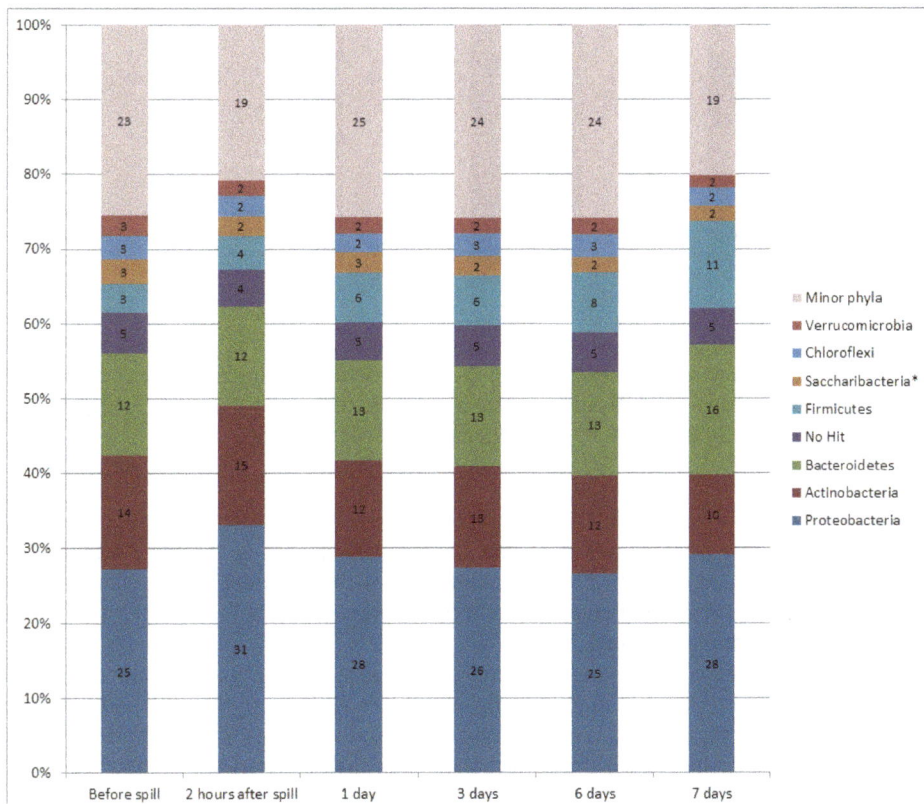

Figure 2:   Bench scale system abundance of OTUs at phylum level. The abundance is presented in terms of percentage from the total number of bacterial sequences in each sample, taxonomical identification was at a confidence threshold of 99% except for "*," which was at 95%. <1% Taxa abundance have been excluded.

Table 2: Chao 1 diversity index.

	Before spill	2 hours after spill	1 day	3 days	6 days	7 days
Chao 1	2009	1838	2000	1900	1995	1948
Shannon	5.295	5.281	5.414	5.324	5.362	5.240

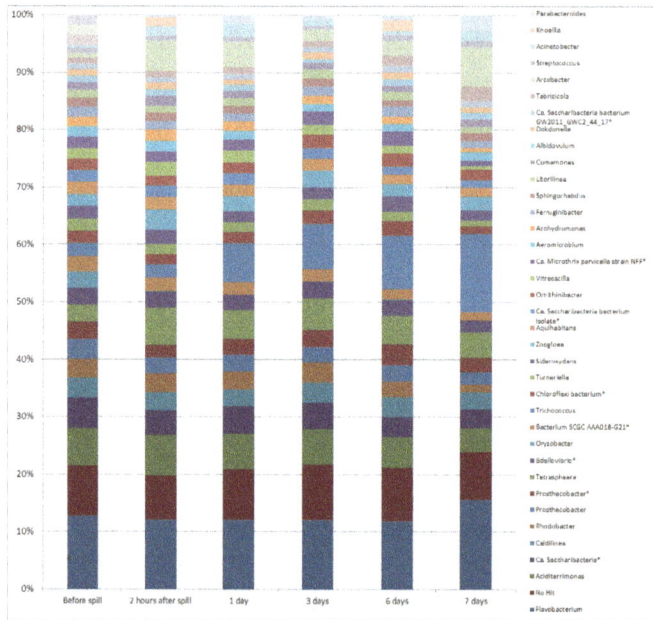

Figure 3: Abundance of OTUs at genera level. The abundance is presented in terms of percentage from the total number of bacterial sequences in each sample, taxonomical identification was at a confidence threshold of 99% except for "*," which was at 95%. <1% abundance have been excluded.

The most abundant genus was the denitrifying Flavobacterium, which has been found as a part of community core in WWTPs and have been reported to produce extracellular polymers that bound cells together, so they could act as floc-forming microorganisms [17]. Zhang et al. [15] also reported Flavobacterium as a dominant genus in 3 WWTP from North America of the 14 samples studied.

The following genera in abundance are Aciditerrimonas, Saccharibacteria and Caldilinea, respectively. Aciditerrimonas is an iron-reducing bacteria as Sideroxydans [18], which is part of the genera core of the activated sludge studied too (abundance >1%). The high abundance of these genera could be related to the fact that both, the activated sludge and the settled influent come from a plant where ferric chloride is used in primary treatment.

Zhang et al. [15] observed that the species composition of biomass varied with geographic location, although microbial genera Zoogloea, Dechloromonas, Prosthecobacter, Caldilinea and Tricoccus existed in all WWTPs. In our case, the same distribution was found, except Dechloromonas, which was absent of the core community, presented relative abundance <1%.

3.2 Genera variation after the spill

Fig. 4 represents the genera that increase at least 40% of their relative abundance after the assay.

Tetrasphaera and Arcobacter are the genera those that grow faster, in the following 2 hours after the spill. Arcobacter is a pathogen bacteria has been observed in wastewater influents [16], [19] and are efficiently eliminated during biological treatment [20]. Besides Trichococcus and Acinetobacter were detected between the 10 genera most abundant in the wastewater influent of 3 WWTPs by Saunders et al. [21]. The increase of Arcobacter, Trichococcus and Acinetobacer could indicate that water treatment biomass efficiency is being affected by the spill, since the relative abundance of these genera increases rapidly at the beginning and also after 6 days from the spill, when the accumulated toxicity effect could be revealed.

Trichococcus is the genus that suffers the greatest increase, specifically represented by the specie *Trichococcus pasteurii*. It must be noted that Saunders et al. [21] underlined that Trichococcus exhibited a high net growth rate in activated sludge beside being abundant in the influent. Zhang et al. [15] identified this genus as psychrotolerant mesophile at high abundances (1.55–5.53%), reaching in this case the 8% in the 7th day from the spill.

Both *Trichococcus pasteurii* and Tetrasphaera genus (mainly *Tetrasphaera elongata* and also *T. vanveenii* and *T. jenkinsii*) have been related with Eikelboom filamentous morphotype *Nostocoida limicola* [22], [23]. This filamentous morphotype causes foaming problems in WWTP. The presence of *Nostocoida limicola*-like organisms was confirmed by conventional microscopy (Fig. 5).

Nostocoida limicola Morphotype has been described as part of the bacterial communities of industrial wastewater treatment plants [24]–[26], suggesting that they could be able to adapt to the presence of toxic substances.

Interestingly, Parabacteroides genus (corresponded to the obligately anaerobic species *Parabacteroides chartae* [27]) increased its relative abundance, while *Tabrizicola aquatica* being aerobic [28], also increased its relative abundance.

In Fig. 6 genera that decrease at least a 40% their relative abundance are shown. Oryzobacter genera suffer a sudden drop only 2 hours after the spill, almost disappearing in the next days (relative abundance <0.016 in day 7). This genus was represented in the samples by the only specie *Oryzobacter terrae*, which is a rod-shaped bacterium isolated from rice soil [29]. This specie seems to be very sensible to the spill effect.

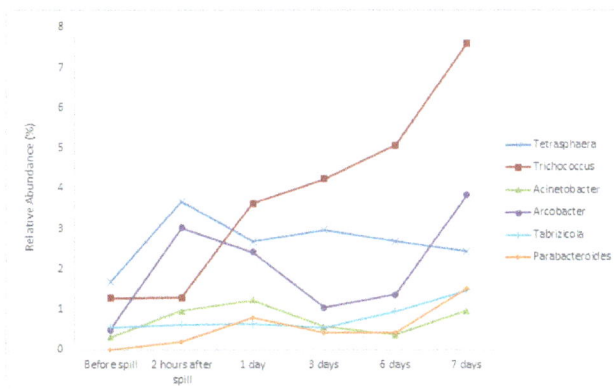

Figure 4: Relative abundance of genera OTUs that increase by at least 40%.

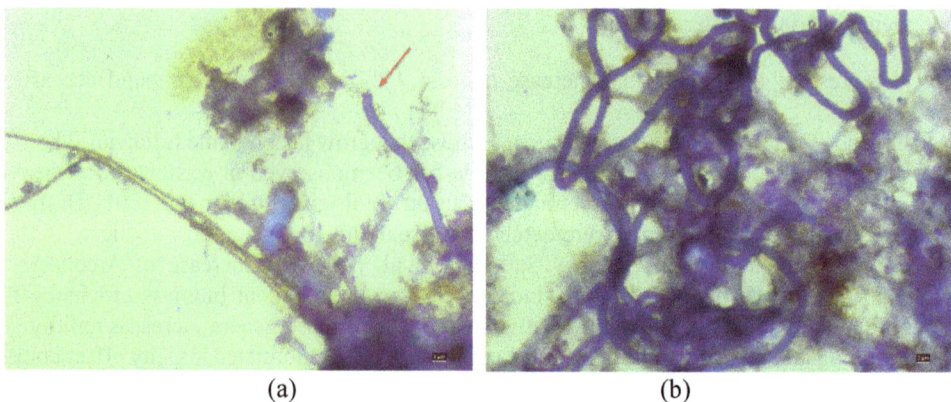

| (a) | (b) |

Figure 5: Positive Neisser stain of N. limicola morphotype. Bright field x1000. (a) Day 2 after the spill; and (b) Day 7 after the spill.

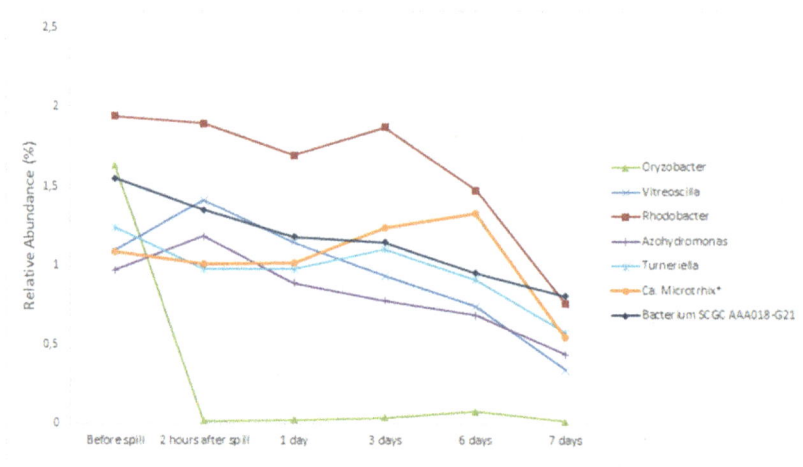

Figure 6: Relative abundance of genera OTUs that decrease by at least 40%.

3.3 Nitrifying genera

Respirometry assays results showed a decrease of 100% of nitrifying activity after the insecticide spill and it was maintained up to the end of the assay. However, ammonia oxidizing bacteria (AOA) and nitrate oxidizing bacteria (NOB) remain almost unchanged during all the study (Fig. 7). Only Rhizobium genera relative abundance, which could carry out nitrite reduction according to Wang et al. [14], varied during the study. This results could indicate that, although nitrifying biomass exists, it was inhibited.

Nitrosomonas and Nitrosospira, are the most important genera of AOB in WWTPs [30], [31]. Nitrosomonas were the most abundant AOB in the assay samples, accounting between 0.16–0.21% of the total sequences of the samples. Low values for nitrifying bacteria were also described by Wang et al. [14]. At the specie level, *Nitrosomonas oligotropha* was the dominant one, the same reported by Limpiyakorn et al. [32].

3.4 Denitrifying genera

Denitrifying bacteria represented in the core community (relative abundance >1% in almost one sample) were, in relative abundance order, Flavobacterium (described as floc-forming [17]), the photosynthetic Rhodobacter, Zoogloea (also floc-forming) and Comamonas. Only Rhodobacter relative abundance decreased over 40% in the assay (Fig. 8), while Flavobacterium increased 20% of its relative abundance at the 7th day.

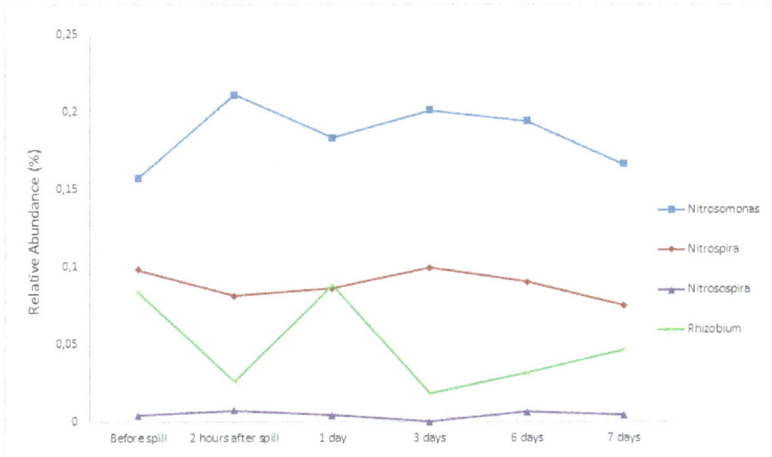

Figure 7: Relative abundance of nitrifying genera OTUs variation after spill (relative abundance <1%).

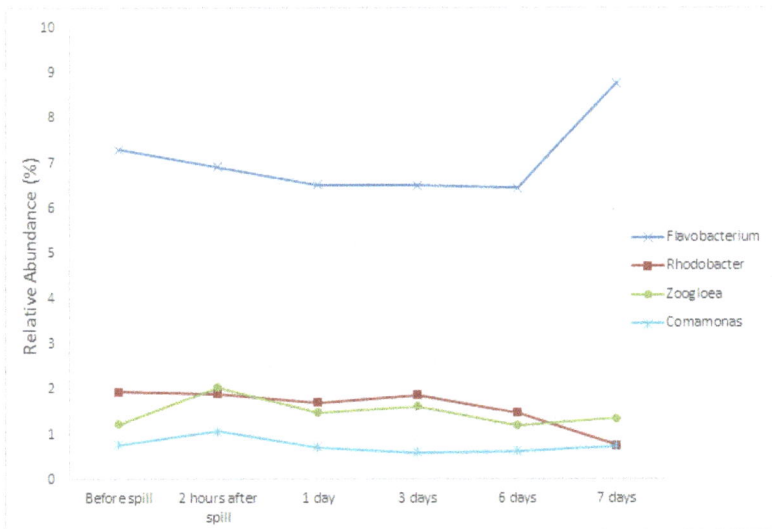

Figure 8: Relative abundance of denitrifying genera OTUs variation (Only OTUs with relative abundance >1%).

4 CONCLUSIONS

The microbial community structure of an activated sludge system at bench-scale before and after an insecticide spill were analyzed by metagenomic analysis. Proteobacteria, Actinobacteria and Bacteroidetes were the most abundant phyla at the beginning of the assay, but the toxic effect in the biomass changed the distribution of the activated sludge population, replacing Firmicutes phylum to Actinobacteria as the 3rd most abundant phylum, while species richness only decreases in short term.

Although nitrifying activity decreased a 100% during the study, nitrifying and denitrifiying genera OTUs have not varied significantly during the assay, except for Rhodobacter, which decreases at the end of the experiment. These results highlight the need of using techniques like the respirometry to complement microbiome analysis, since the presence of certain genera do not ensure that they are active in the biomass.

Arcobacter, Trichococcus and Acinetobacer genera, identified as abundant in water inlet, increased its relative abundance after the spill, which might evidence an affection in the purification performance capacity of the biomass. On the other hand, Oryzobacter genus, represented by *Oryzobacter terrae*, appears as very sensitive to the toxic effect of spill almost disappearing after 2 hours of the insecticide exposition. Taking into account that Trichococcus can grow in the activated sludge, that results indicate that Arcobacter, Acinetobacer and Oryzobacter are good candidates to be monitored as indicators of a negative affection in the activated sludge biological process studied. However more studies are required to define the potential use of these genera to develop a system based in "omic" techniques to early detect malfunctions in the activated sludge process by alteration of the microbial populations.

ACKNOWLEDGEMENTS

The authors were grateful for the financial support from the European Commission through LIFE Programme (LIFE ENV16/ES/000390). The opinions or points of view published herein do not represent EC official position.

REFERENCES

[1]   Zhu, A., Guo, J., Ni, B-J., Wang, S., Yang, Q. & Peng, Y., A novel protocol for model calibration in biological wastewater treatment. *Nature Scientific Reports*, **5**, p. 8493, 2015.

[2]   Vanwonterghem, I., Jensen, P.D., Ho, D.P., Batstone, D.J. & Tyson G.W., Linking microbial community structure, interactions and function in anaerobic digesters using new molecular techniques. *Current Opinion in Biotechnology*, **27**, pp. 55–64, 2014. https://doi.org/10.1016/j.copbio.2013.11.004.

[3]   Cydzik-Kwiatkowska, A. & Zielin´ska, M., Bacterial communities in full-scale wastewater treatment systems. *World Journal of Microbiology and Biotechnology*, **32**(4), pp. 66, 2016. https://doi.org/10.1007/s11274-016-2012-9.

[4]   Environmental cost-effective activation treatment for biological failures in Wastewater Treatment Plants. https://www.bactiwater.com/en/.

[5]   Klindworth, A. et al., Evaluation of general 16S ribosomal RNA gene PCR primers for classical and next-generation sequencing-based diversity studies. *Nucleic Acids Research*, **41**(1), p. e1, 2013.

[6]   Zhang, J., Kobert, K., Flouri, T. & Stamatakis, A., PEAR: A fast and accurate Illumina Paired-End read mergeR. *Bioinformatics,* **30**, pp. 614–620, 2014.

[7]   Martin, M., Cutadapt removes adapter sequences from high-throughput sequencing reads. *EMBnet.journal*, **17**, pp. 10–12, 2011.

[8]   Altschul, S.F., Gish, W., Miller, W., Myers, E.W. & Lipman, D.J., Basic local alignment search tool. *Journal of Molecular Biology*, **215**, pp. 403–10, 1990.

[9]   Meerbergen, K. et al., Assessing the composition of microbial communities in textile wastewater treatment plants in comparison with municipal wastewater treatment plants. *MicrobiologyOpen*, **6**(1), 2017. https://doi.org/10.1002/mbo3.413.

[10] Liang, H., Ye, D. & Luo, L., Unravelling diversity and metabolic potential of microbial consortia at each stage of leather sewage treatment. *RSC Advances Journal*, **7**(66), pp. 41727–41737, 2017. https://doi.org/10.1039/C7RA07470K.

[11] Yang, Y. et al., Metagenomic analysis of sludge from full-scale anaerobic digesters operated in municipal wastewater treatment plants. *Applied Microbiology and Biotechnology*, **98**, 5709–5718, 2014. https://doi.org/10.1007/s00253-014-5648-0.

[12] Liu, C., Li, H., Zhang, Y., Si, D. & Chen, Q., Evolution of microbial community along with increasing solid concentration during high-solids anaerobic digestion of sewage sludge. *Bioresource Technology*, **216**, 87–94, 2016. https://doi.org/10.1016/J.BIORTECH.2016.05.048.

[13] Yi, J., Dong, B., Jin, J. & Dai, X., Effect of increasing total solids contents on anaerobic digestion of food waste under mesophilic conditions: Performance and microbial characteristics analysis. *PLoS One*, **9**, p. e102548, 2014. https://doi.org/10.1371/journal.pone.0102548.

[14] Wang, Z. et al., Abundance and Diversity of Bacterial Nitrifiers and Denitrifiers and Their Functional Genes in Tannery Wastewater Treatment Plants Revealed by High-Throughput Sequencing. *PLoS ONE*, **9**(11), p. e113603, 2014. https://doi.org/10.1371/journal.pone. 0113603.

[15] Zhang, T., Shao, M.F. & Ye, L., 454 pyrosequencing reveals bacterial diversity of activated sludge from 14 sewage treatment plants. *ISME Journal*, **6**(6), pp. 1137–1147, 2012.

[16] Gonzalez-Martinez, A. et al., Comparison of bacterial communities of conventional and A-stage activated sludge systems. *Scientific Re*ports, **6**, p. 18786, 2016. https://doi.org/10.1038/srep18786.

[17] Guo, F., Zhang, S.-H., Yu, X. & Wei, B., Variations of both bacterial community and extracellular polymers: The inducements of increase of cell hydrophobicity from biofloc to aerobic granule sludge. *Bioresource Technology*, **102**, pp. 6421–6428, 2011.

[18] Shivlata, L. & Satyanarayana, T., Thermophilic and alkaliphilic Actinobacteria: biology and potential applications. *Frontiers Microbiology*, 2015. https://doi.org/10.3389/fmicb.2015.01014.

[19] Stampi, S., Varoli, O., Zanetti, F., & De Luca, G., Arcobacter cryaerophilus and thermophilic campylobacters in a sewage treatment plant in Italy: Two secondary treatments compared. *Epidemiology and Infection*, **110**(3), pp. 633–639, 1993. https://doi.org/10.1017/S0950268800051050.

[20] Xin, L. et al., Bacterial pathogens and community composition in advanced sewage treatment systems revealed by metagenomics analysis based on high-throughput sequencing. *PLoS One*, **10**(5), p. e0125549, 2015. https://doi.org/10.1371/journal.pone.0125549.

[21] Saunders, A.M., Albertsen, M., Vollertsen, J. & Nielsen, P.H., The activated sludge ecosystem contains a core community of abundant organisms. *The ISME Journal*, **10**, pp. 11–20, 2016.

[22] Liu, J.R. et al., Emended description of the genus Trichococcus, description of Trichococcus collinsii sp. nov., and reclassification of Lactosphaera pasteurii as Trichococcus pasteurii comb. nov. and of Ruminococcus palustris as Trichococcus

palustris comb. nov. in the low-G+C gram-positive. *International Journal of Systematic and Evolutionary Microbiology,* **52**(4), 2002.

[23]   Seviour, R. and Nielsen P.H., *Microbial Ecology of Activated Sludge,* IWA Publishing, p. 307, 2010.

[24]   Rodríguez, E., Isac, L., Fernández, N., Zornoza, A. & Mas, M., Identificación de bacterias filamentosas en EDAR industriales. *Tecnología del Agua,* **303**, pp. 56–64, 2008.

[25]   Eikelboom, D.H. & Geurkink, B., Filamentous micro-organisms observed in industrial activated sludge plants. *Water Science and Technology,* **46**, pp. 535–542, 2002.

[26]   Eikelboom, D.H., *Identification and Control of Filamentous Microorganisms in Industrial Wastewater Treatment Plants,* IWA Publishing: London, 2006.

[27]   Tan, H.Q., Li, T.T., Zhu, C., Zhang, X.Q., Wu, M. & Zhu, X.F., Parabacteroides chartae sp. nov., an obligately anaerobic species from wastewater of a paper mill. *International Journal of Systematic and Evolutionary Microbiology,* **62**(11), pp. 2613–2617, 2011. https://doi.org/10.1099/ijs.0.038000-0.

[28]   Tarhriz, V., Thiel, V., Nematzadeh, G., Hejazi, M.A., Imhoff, J.F. & Hejazi, M.S., Tabrizicola aquatica gen. nov. sp. nov., a novel alphaproteobacterium isolated from Qurugöl Lake nearby Tabriz city, Iran. *Antonie van Leeuwenhoek,* **104**(6), pp. 1205–15, 2013. https://doi.org/10.1007/s10482-013-0042-y.

[29]   Kim, S.J. et al., Oryzobacter terrae gen. nov., sp. nov., isolated from paddy soil. *International Journal of Systematic and Evolutionary Microbiology,* **65**(9), pp. 3190–3195, 2015. https://doi.org/10.1099/ijsem.0.000398.

[30]   Park, H., & Noguera, D., Evaluating the effect of dissolved oxygen on ammonia-oxidizing bacterial communities in activated sludge. *Water Research,* **38**(14–15), pp. 3275–3286, 2004. https://doi.org/10.1016/j.watres.2004.04.047.

[31]   Purkhold, U., Pommerening-Roser, A., Juretschko, S., Schmid, M.C., Koops, H.P. & Wagner, M., Phylogeny of all recognized species of ammonia oxidizers based on comparative 16S rRNA and amoA sequence analysis: implications for molecular diversity surveys. *Applied and Environmental Microbiology,* **66**(12), pp. 5368–5382, 2000. https://doi.org/10.1128/AEM.66.12.5368-5382.2000.

[32]   Limpiyakorn, T., Kurisu, F. & Yagi, O., Quantification of ammoniaoxidizing bacteria populations in full-scale sewage activated sludge systems and assessment of system variables affecting their performance. *Water Science and Technology,* **54**(1), pp. 91–99, 2006.

EFFECT OF LANDSCAPE METRICS ON WATER QUALITY OVER THREE DECADES: A CASE STUDY OF THE AVE RIVER BASIN, PORTUGAL

ANTÓNIO CARLOS PINHEIRO FERNANDES[1], LISA MARIA DE OLIVEIRA MARTINS[1],
LUÍS FILIPE SANCHES FERNANDES[1] & FERNANDO ANTÓNIO LEAL PACHECO[2]
[1]CITAB – Centre for the Research and Technology of Agro-Environment and Biological Sciences,
University of Trás-os-Montes and Alto Douro, Portugal
[2]CQVR – Vila Real Chemistry Research Centre, University of Trás-os-Montes and Alto Douro, Portugal

ABSTRACT

Due to intense industrial and urban activity Ave River Basin, Portugal was once tagged as one of the most polluted in Europe. Besides point source pressures are the most evident threat to water quality in this river basin, in the present study, the effect of landscape on water quality was analysed. From a hydrological database, the concentration surface water parameters was extracted, comprehended between 1988 and 2016. The average concentrations in each sampling site for each hydrological year was calculated. The averages were correlated to 15 landscape metrics by using the Spearman's rank correlation coefficient, over the analysed hydrological years. For each landscape metric, the percentage of correlations with surface water parameters for each hydrological year that had statistical significance ($p \leq 0.05$) was counted, and the same analysis was made for each surface water parameter. The area occupied by artificial surfaces increased the contaminant concentrations in 65.3% of the correlations while the edge density increased by 62%. For forested areas, the edge density and occupied area had, respectively, 56.0% and 66.0%, of correlations that decreased contaminant concentrations. Conductivity was the parameter that has most linked to landscape metrics since, 52% increased the concentration, 21% decreased, while the remaining 27% did not have statistical significance. Oxygen demands, total suspended solids, different nitrates forms, total orthophosphate and coliforms were acceptably correlated, with percentages ranging from 20% to 44%. Only heavy metals were poorly correlated, since the percentage of correlations that varied the concentrations was lower than 8%. This study allowed us to understand that in an urbanised river basin, where point source pressures are the dominant pollution source, landscape metrics also have an effect on water quality and can become a threat to hydric resources.
Keywords: Spearman's rank correlation coefficient, water quality, ArcGIS, Ave River Basin, landscape metrics.

1 INTRODUCTION

The management of hydric resources is one challenge that has alongside humanity for thousands of years. Not only the availability of hydric resources but also the quality is a concern that emerged with demographic expansion and economic development. Since many threats to water quality have been arising, researchers have been conducting their studies in the scope of improving treatment technologies [1], prevention and mitigation strategies [2]. To study water quality (WQ) is necessary to comprehend that there is a multitude of interactions [3]–[7] and a vast type of different pollution sources that will end up releasing contaminants in hydric resources [8], which is essential knowledge to support the management of hydric resources [9]. In general, two types of pollution sources exist, point source (PS) and non-point source (NPS). Effluent discharges are the clearest point source pressures in hydric resources [10]. Besides the purpose of treatment stations is to reduce contamination loads, when treatment stations are not under proper functioning hydric resources become heavily contaminated. Commonly, effluents from domestic sewage contain high loads of organic compounds [11], while in industrial effluents, the composition is highly

WIT Transactions on Ecology and the Environment, Vol 242, © 2020 WIT Press
www.witpress.com, ISSN 1743-3541 (on-line)
doi:10.2495/WP200041

variable. Metallurgic and textile industries may release heavy metals [12], [13], pharmaceutical industries can release different chemical compounds [14], and food industries can release high loads of organic compounds [15], [16]. The diffuse sources (also called NPS) are large areas that can contain contaminants that are carried to rivers through surface runoff mechanisms. Such areas might be agricultural areas or livestock farms [17]. In agricultural fields, the pollution comes from the application of herbicides and pesticides [18], and in livestock areas, manure is a contamination source [19]. Many studies have been executed in the scope of the impact of NPS in WQ. In such studies, landscape metrics have been used as variables that characterise diffuse pollution sources. The composition of a landscape allows to interpret which are the dominant land uses, but the configuration,(for example, edge density and number of patches) are also key aspects to interpret if a landscape is vulnerable or not to diffuse pressures [20].

The present study takes place in an urbanised river basin located in the northwest region of Portugal, Ave River Basin (Fig. 1). Due to intense anthropogenic activity, during the second half of the 20th century, industrial effluent discharges without proper treatment. For such reasons, it became one of the most polluted in Europe [21]. This problematic caught the attention of several experts to study this problem. In pioneer studies, heavy metals concentration was measured in surface waters, aquatic fauna and sediments, to evidence the problem [13], [22]. Due to an ambitions remediation program, the concentrations dropped in the begging of the 21st century [23]. However, in more recent studies, it was revealed that PS was still contaminating the river basin and also diffuse pressures became another concern [24]–[26].

The effects of landscape metrics on surface water parameters from 1988 to 2016 in Ave River Basin were studied, with the purpose of understanding which contaminants are related to landscape, and which landscape metrics are the one threat to WQ.

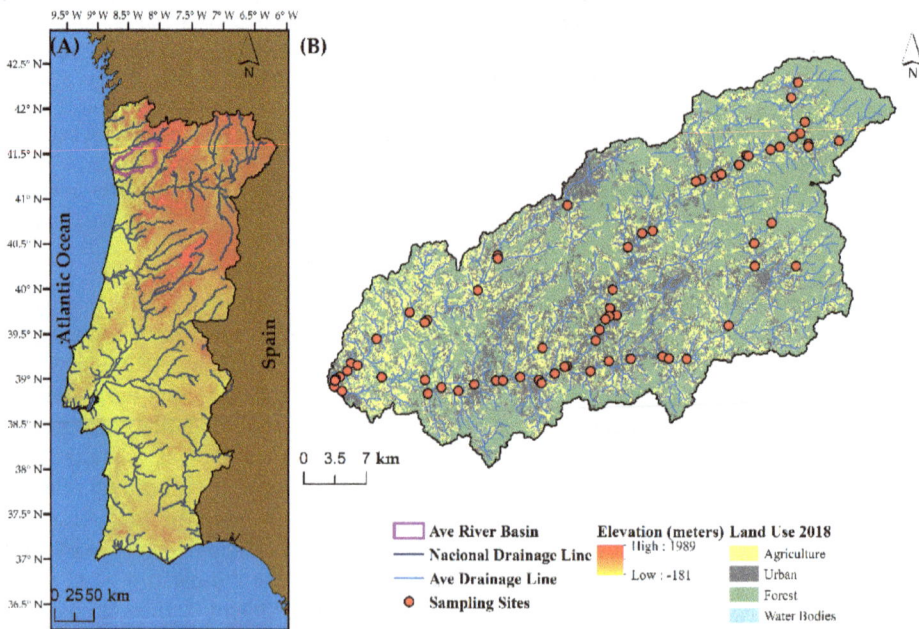

Figure 1: Ave River Basin location. (a) Portugal; and (b) Ave River Basin land use of 2018, and location of SWP sampling sites.

2 METHODOLOGY

The relation between landscape metrics and surface water parameters along different hydrological years (HY) was analysed by using Spearman's correlation rank coefficient. The first step was to delineate the Ave River Basin, drainage lines and also the drainage area of each WQ sampling site, by using ArcMap [27] and ArcHydro tools [28]. The location of the sampling sites and also the measurements of different surface water parameters (SWP) was downloaded from the Portuguese hydrological database SNIRH [29]. Table 1 represents the number of sampling sites for each SWP according to the HY.

The land cover was downloaded from the Portuguese Territorial database. For each year and drainage area, it was calculated landscape metrics by using a python toolbox [30]. Metrics were calculated for generic land uses, agriculture areas (AGR), artificial surfaces (ART) and also forest and semi-natural areas (FOR). The calculated metrics were the number of patches (npc), the percentage of area occupied (pz), the edge density (ed) and also the percentage of edges that are shared with two different types of land use (cce).

The Spearman's correlation rank coefficient (r_s), measures monotonic relation between the paired variables. This coefficient varies from -1 to 1. When r_s is close to 0, the correlation is weak, while if the value is closer to -1 or 1 it means that the correlation is strong. In order to know if a correlation is statistically significant, is crucial to analyse the probability to reject the null hypothesis, which is dependent on the sample size. For two-tailed probabilities and a statistical significance of 0.05, the minimum sample size to achieve a statistically significant correlation is 5. Since for that sample size is necessary to have the maximum correlation ($r_s=1$ or $r_s=-1$) it was analysed only the correlations with at least six samples.

3 RESULTS AND DISCUSSION

For the present study, it was calculated the correlations between the 15 LSM with 14 SWP for the 29 analysed hydrological years. The correlations were counted, and it was calculated the percentage of correlations that had statistical significance. Tables 2 and 3 present the positive and negative correlations, respectively. For the percentage calculation, it was considered only correlations that had six or more samples. In Tables 2 and 3, the centred cells represent the percentage of hydrological years were the correlation was statistically significant. The last columns are the percentage of hydrological years among all metrics for the respective SWP, while in the last row is made the same analysis but for each LSM. For example, conductivity is a parameter that had more than five samples in 27 hydrological years. By looking to the percentage of correlations between conductivity with npc_(ART) in Table 2, the percentage of correlations is approximately 59.3%, which means that in 16 of the 27 analysed hydrological years, the correlation between this two variables is positive and statistically significant. In the last row of Table 2, the percentage of correlations among all LSM with conductivity is approximately 51.9%. Which means that in a total of 405 correlations (15 LSM in 27 HY), conductivity was positively correlated with statistical significance in 210 correlations, while 85 of the correlations (21% in Table 3) were negative and statistically significant, but the remaining 110 correlations were not statistically significant. By analysing the percentage of correlations of npc_(ART) in Table 2, is seen that the percentage of correlations is approximately 45.1%. This percentage is calculated by the total number of correlations among all SWP in the available hydrological years that had more than five measurements which is in total 297 correlations, and 45.1% of the total number of calculated correlations is 134.

Table 1: Surface water parameters and the number of sampling sites with data.

Hydrological year	Total suspended solids	Conductivity	Biological oxygen demand	Chemical oxygen demand	Total nitrate	Ammonium	Ammonia	Orthophosphate	Fecal coliforms	Total coliforms	Chromium	Lead	Cadmium	Mercury
1988–1989	3	3	3	0	0	3	0	0	3	0	0	0	0	0
1989–1990	6	6	6	6	6	6	0	6	6	6	0	0	0	0
1990–1991	6	6	6	6	6	6	1	6	6	6	5	5	5	5
1991–1992	6	6	6	6	6	6	0	6	6	0	0	0	0	0
1992–1993	1	3	1	3	3	3	0	3	3	2	0	0	0	0
1993–1994	15	15	15	15	15	15	0	15	15	9	0	0	0	0
1994–1995	15	15	15	15	15	15	15	15	15	9	0	0	0	0
1995–1996	15	15	15	15	15	15	15	15	15	9	0	0	0	0
1996–1997	16	17	16	17	17	17	17	17	17	11	0	0	0	0
1997–1998	17	17	17	17	17	17	17	17	17	11	0	0	0	0
1998–1999	20	20	20	20	20	20	20	20	20	11	2	2	2	2
1999–2000	18	18	18	17	18	18	19	16	17	6	6	6	6	6
2000–2001	10	7	10	7	10	10	11	7	7	7	5	5	5	5
2001–2002	11	9	9	8	8	9	10	8	8	8	9	11	11	11
2002–2003	12	9	9	8	9	9	10	8	8	8	9	9	9	9
2003–2004	12	9	9	8	9	9	10	8	8	8	9	9	9	9
2004–2005	13	13	13	12	12	13	0	12	12	12	12	12	12	12
2005–2006	13	13	13	12	12	13	0	12	12	12	12	12	12	12
2006–2007	14	14	14	13	14	14	0	13	12	12	12	12	12	12
2007–2008	14	14	14	12	12	14	14	12	11	11	11	0	0	0
2008–2009	27	24	23	22	28	28	24	22	21	21	21	0	0	0
2009–2010	31	31	31	30	30	31	23	25	23	23	23	14	14	14
2010–2011	20	21	22	19	23	23	5	16	14	14	0	0	0	0
2011–2012	19	18	19	16	16	17	5	16	14	14	16	16	16	15
2012–2013	17	17	17	17	17	17	5	16	0	0	0	0	0	0
2013–2014	28	28	28	15	28	28	5	13	0	0	13	12	13	13
2014–2015	29	29	29	8	29	29	0	6	0	0	0	0	0	0
2015–2016	8	13	9	8	9	8	0	6	0	0	0	0	0	0
2016–2017	8	8	8	0	8	8	0	0	0	0	0	0	0	0

By comparing Table 2 with Table 3, is seen in overall that landscape metrics have a stronger influence in the increase of SWP rather than in the decrease. For example, the percentage of correlations that increase conductivity (51.9%) is lower than the percentage that decreases it (21.0%), and in other SWP the percentages of positive correlations are also higher. According to Table 2 (last column), conductivity is mostly increased by landscape

Table 2: Percentage of positive correlations that had statistical significance.

	npc_(AGR)	npc_(ART)	npc_(FOR)	pz_(AGR)	pz_(ART)	pz_(FOR)	cce_(ART)_with_(AGR)	cce_(FOR)_with_(AGR)	cce_(AGR)_with_(ART)	cce_(FOR)_with_(ART)	cce_(AGR)_with_(FOR)	cce_(ART)_with_(FOR)	ed_(AGR)	ed_(ART)	ed_(FOR)	
Total suspended solids	44.4	51.9	22.2	59.3	66.7	0.0	63.0	0.0	0.0	0.0	22.2	66.7	63.0	66.7	0.0	35.1
Conductivity	88.9	88.9	59.3	77.8	88.9	0.0	88.9	0.0	0.0	0.0	25.9	85.2	85.2	88.9	0.0	51.9
Biological oxygen demand	48.1	48.1	37.0	55.6	77.8	0.0	77.8	0.0	0.0	0.0	14.8	63.0	70.4	74.1	0.0	37.8
Chemical oxygen demand	42.3	53.8	23.1	69.2	73.1	0.0	76.9	0.0	0.0	0.0	15.4	65.4	65.4	73.1	0.0	37.2
Total nitrate	25.9	40.7	18.5	85.2	88.9	0.0	88.9	0.0	0.0	0.0	40.7	88.9	96.3	88.9	0.0	44.2
Ammonium	70.4	70.4	40.7	55.6	77.8	0.0	77.8	0.0	0.0	0.0	14.8	74.1	74.1	77.8	0.0	42.2
Ammonia	23.5	23.5	23.5	11.8	52.9	0.0	52.9	0.0	0.0	0.0	0.0	41.2	35.3	41.2	0.0	20.4
Total orthophosphate	69.2	73.1	38.5	76.9	80.8	0.0	88.5	0.0	0.0	0.0	7.7	73.1	73.1	76.9	0.0	43.8
Fecal coliforms	18.2	22.7	9.1	59.1	77.3	0.0	63.6	0.0	0.0	0.0	0.0	63.6	59.1	68.2	0.0	29.4
Total coliforms	28.6	38.1	33.3	47.6	81.0	0.0	66.7	0.0	0.0	0.0	4.8	42.9	38.1	57.1	0.0	29.2
Chromium	14.3	14.3	7.1	7.1	14.3	0.0	14.3	0.0	0.0	7.1	0.0	0.0	14.3	14.3	0.0	7.1
Lead	0.0	0.0	0.0	0.0	8.3	8.3	8.3	8.3	0.0	16.7	0.0	0.0	0.0	16.7	8.3	5.0
Cadmium	0.0	0.0	8.3	0.0	0.0	8.3	0.0	8.3	0.0	16.7	0.0	0.0	0.0	0.0	8.3	2.8
Mercury	8.3	8.3	8.3	0.0	0.0	8.3	0.0	8.3	0.0	16.7	0.0	0.0	0.0	0.0	8.3	4.4
	40.7	45.1	26.6	51.9	65.3	1.0	64.0	1.0	0.0	2.4	13.1	56.6	57.2	62.0	1.0	

Table 3: Percentage of negative correlations that had statistical significance.

	npc_(AGR)	npc_(ART)	npc_(FOR)	pz_(AGR)	pz_(ART)	pz_(FOR)	cce_(ART)_with_(AGR)	cce_(FOR)_with_(AGR)	cce_(AGR)_with_(ART)	cce_(FOR)_with_(ART)	cce_(AGR)_with_(FOR)	cce_(ART)_with_(FOR)	ed_(AGR)	ed_(ART)	ed_(FOR)	
Total suspended solids	0.0	0.0	0.0	0.0	0.0	74.1	0.0	59.3	37.0	7.4	0.0	0.0	0.0	0.0	66.7	16.3
Conductivity	0.0	0.0	0.0	0.0	0.0	88.9	0.0	81.5	51.9	7.4	0.0	0.0	0.0	0.0	85.2	21.0
Biological oxygen demand	0.0	0.0	0.0	0.0	0.0	77.8	0.0	74.1	48.1	7.4	0.0	0.0	0.0	0.0	66.7	18.3
Chemical oxygen demand	0.0	0.0	0.0	0.0	0.0	73.1	0.0	61.5	23.1	7.7	0.0	0.0	0.0	0.0	61.5	15.1
Total nitrate	0.0	0.0	0.0	0.0	0.0	88.9	0.0	85.2	40.7	14.8	0.0	0.0	0.0	0.0	96.3	21.7
Ammonium	0.0	0.0	0.0	0.0	0.0	77.8	0.0	74.1	51.9	11.1	0.0	0.0	0.0	0.0	63.0	18.5
Ammonia	0.0	0.0	0.0	0.0	0.0	47.1	0.0	58.8	11.8	17.6	0.0	0.0	0.0	0.0	47.1	12.2
Total orthophosphate	0.0	0.0	0.0	0.0	0.0	80.8	0.0	73.1	34.6	0.0	0.0	0.0	0.0	0.0	76.9	17.7
Fecal coliforms	0.0	0.0	0.0	0.0	0.0	77.3	0.0	68.2	22.7	13.6	0.0	0.0	0.0	0.0	50.0	15.5
Total coliforms	0.0	0.0	0.0	0.0	0.0	85.7	0.0	52.4	33.3	9.5	0.0	0.0	0.0	0.0	33.3	14.3
Chromium	0.0	0.0	0.0	0.0	0.0	14.3	0.0	7.1	7.1	0.0	0.0	0.0	0.0	0.0	7.1	2.4
Lead	8.3	8.3	8.3	8.3	8.3	8.3	8.3	8.3	0.0	0.0	0.0	8.3	8.3	8.3	8.3	6.7
Cadmium	8.3	8.3	8.3	8.3	8.3	0.0	8.3	0.0	0.0	0.0	0.0	8.3	8.3	8.3	0.0	5.0
Mercury	8.3	8.3	8.3	8.3	8.3	0.0	8.3	0.0	8.3	0.0	0.0	8.3	8.3	8.3	0.0	5.6
	1.0	1.0	1.0	1.0	1.0	66.0	1.0	58.6	31.3	7.7	0.0	1.0	1.0	1.0	55.9	

metrics 51.9%, followed by parameters related to nutrients (N and P), nitrate (44.2%), total orthophosphate (43.8%) and also ammonium with (42.2%). With lower percentages, oxygen demands, both BOD_5 and COD are still affected by landscape metrics with a percentage of 37.8 and 37.2%, respectively. Total suspended solids percentage is 35.1%, coliforms are also increased in 29.4% and 29.2, faecal and total respectively. Ammonia is the parameter related to N forms that is less influenced by landscape metrics with a percentage of 20.4%, and heavy metals are less correlated to LSM since the percentage of positive correlations are the lowest, ranging from 7.1% to 2.8%.

The parameters that are mostly decreased by LSM (Table 3 last column) are total nitrate (21.7%) and conductivity (21.0%). Other parameters have percentages that vary from 18.5% to 12.2%, expect heavy metals that have lower percentages, ranging from 6.7% and 2.4%.

In the last row of Tables 2 and 3 the percentage of correlations associated with each metric is shown. Some contrast metrics might look identical when applied to the same pair of land uses, for example, cce_(AGR)_with_(FOR) and cce_(FOR)_with_(AGR). However, they are different since in the first case is the percentage of forested edges that are shared with agriculture edges, while the second metric is the percentage of agricultural edges that are shared with forestry edges. From the 15 LSM only four have a notorious impact in the decrease of contaminants concentrations, pz_(FOR) (66.0%), cce_(FOR)_with_(AGR) (58.6%), ed_(FOR) (55.9%) and unexpectedly the cce_(AGR)_with_(ART) (31.3%). There are eight metrics that increase the SWP: pz_(ART) (65.3%), cce_(ART)_with_(AGR) (64%), ed_(ART) (62%), ed_(AGR) (57.2%), cce_(ART)_with_(FOR) (56.6%), pz_(AGR) (51.9%), npc_(ART) (45.1%) and npc_(AGR) (40.7%). The remaining three metrics, npc_(FOR), cce_(AGR)_with_(FOR) and cce_(FOR)_with_(ART), can be assumed as non-effective since the percentages that increase and decrease SWP are much lower when compared to the other LSM.

In the main matrix of Tables 2 and 3, it can be seen in detail which SWP are related to LSM. For heavy metals the percentages of correlations are quite low, varying from 16.7% to 0.0%, so there is not any particular LSM that can be related to heavy metals. The npc_(AGR) and npc_(ART) are both related to the increase of conductivity, orthophosphate and ammonium. The pz_(AGR) is linked to SWP with the expectation of ammonia and heavy metals, while the pz_(ART) and cce_(ART)_with_(AGR) are not linked to heavy metals. The cce_(ART)_with_(FOR) and ed_(AGR) are less linked to ammonia and total coliforms, and not linked to heavy metals, while ed_(ART) is more linked to the increase of total coliforms.

The pz_(FOR) is the metric that is mostly related to the decrease of SWP, and it decreases all SWP (except heavy metals) since the percentage of correlations ranges from 88.9–47.1%. The cce_(FOR)_with_(AGR) also decreases the SWP but with a shorter range of 85.2–52.4% (except heavy metals). The cce_(AGR)_with_(ART) decreases, in general, the SWP but seen in detail there are SWP that are more related to this metric, specific conductivity and ammonium both with 51.9% and BOD with 48.1%. The ed_(FOR) is another LSM that had a strong influence in the decrease of SWP. However, it has high variability in the percentages of particular SWP, since in the decrease of nitrates the percentage is 96.3%, for conductivity is 82.5%, orthophosphate is 76.9%, and for the reaming SWP is lower than 70%, and also has no influence in heavy metals.

4 DISCUSSION

The used methodology allowed to understand which are the effects of landscape metrics in Ave River Basin. The analysis consisted in summarize the correlations between LSM and SWP in during many hydrological years, in order to treat the effects for an extended period.

In general, the metrics have a substantial effect on the increase of contaminant concentrations rather than in the decrease. This is an expected result since there were used only three types of land use, of which ART and AGR usually have a negative impact on WQ, while FOR increases WQ [31].

This study allowed to understand which SWP are associated to LSM. By summing the respective percentages of Tables 2 and 3, is calculated the total percentage of statistically significant correlations. Therefore, conductivity and total nitrates are the variables that are more influenced by landscape metrics, since the sum of percentages is the highest, respectively 72.8% and 65.9%. Since electrical conductivity is a measure of ions is a parameter that portrays many contaminants surface water, which are linked to landscape metrics. The presence of nutrients is commonly accessed in landscape metrics studies [32], and also, in this case, the concentration of nitrates is linked to landscape metrics, and also total orthophosphate, which is one of the phosphorous. Oxygen demands are also linked in other studies with landscape metrics [33]–[35], in the present study BOD (56.1%) and COD (52.3%) are linked to LSM. The total suspended solids is also another parameter that is related to LSM, since the total percentage is 51.4%, while the others have percentages lower than 50%, possibly because they are more linked to point source pressures, which reflects in low correlations with LSM [36]. Total coliforms and faecal have total percentages of 43.5% and 44.9%, respectively, besides this contaminant can be from livestock origin, it can be more linked to domestic sewage [37]. Ammonia can be released into surface water from different sources, such as fertilizer and also from industrial applications [38]. Since the total percentage is 32.6%, it can be assumed that in the Ave River Basin is more linked to point source pressures rather than to diffuse emissions. The total percentage of the studied heavy metals ranges from 11.7% (lead) to 7.8% (cadmium), which means that the presence of this SWP is intrinsically linked to point sources pressures, as other studies have revealed for the studied river basin [13], [22], [39].

The results led to understand that the landscape metrics that play a dominant role in the river basin WQ are, the pz_(FOR) and pz_(ART), since pz_(FOR) decreases the concentration of SWP in 66% of the correlations while pz_(ART) increases the SWP in 65.9%. Clearly, these variables expose the effect of natural areas vs anthropogenic regions, because in forestry areas found freshwaters with high quality [40]. Besides, urban areas can also be linked to point source pressures, urban areas are impervious, and the drainage can be routed surface waters, which is a form of diffuse urban pressure [41]. Besides the composition of artificial surfaces and forestry areas has an impact in Ave River Basin, other metrics evidenced influence on SWP, and also the ed_(ART) increases SWP with a percentage of positive correlations of 62.0%. The percentage of agricultural edges that are shared with artificial surfaces, cce_(ART)_with_(AGR) has a strong impact, since the percentage of correlations that decrease WQ is 64.0%. This was expected since agricultural areas, and artificial surfaces are land use types that commonly decrease WQ [31]. On the other hand, the cce_(AGR)_with_(ART) revealed 31.3% negative correlations which is hard to explain. Separately the ed_(AGR) also revealed a degradation effect with 57.2% positive correlations. Nevertheless, the percentage of agricultural edges that are shared with forestry, cce_(FOR)_with_(AGR), revealed a positive impact on WQ, since the percentage of negative correlations (58.6%). It is curious that edge density of forests, ed_(FOR), has a slightly lower percentage, (55.9%). Other authors have reached a similar result [42], and this shows that agricultural fields that are surrounded by forests might not be harmful to hydric resources, by function as sinks for the contaminant flow. However, the percentage of forested areas that are shared with urban areas cce_(ART)_with_(FOR), has a negative impact on WQ, possibly because the effect of urban areas overcomes forestry regions.

5 CONCLUSION
The results, clearly evidenced that the relations between LSM and SWP is highly variable depending on the metric and the parameter. The results clearly showed that in Ave River Basin, urban areas and also agricultural surfaces degrade WQ, while forestry land use can improve WQ. In practical terms, is necessary to increase forested areas in spread shapes, increasing edge density and total area. While for urban areas besides, it is hard to decrease the area occupied, is necessary to at least reduce the urban sprawl. In agricultural areas is essential to surround them with forests, in order to promote a natural barrier for the contaminant flow. For the biggest part of the analysed SWP, conscious land use changes might reduce the contamination into sustainable values. However, to reduce the contamination of ammonia and heavy metals is necessary to implement measures that are beyond landscape metrics, which can be the implementation of improved effluent treatment technologies.

ACKNOWLEDGEMENTS
This research was funded by the INTERACT project – "Integrated Research Environment, Agro-Chain and Technology", no. NORTE-01-0145-FEDER-000017, in its line of research entitled BEST, co-financed by the European Regional Development Fund (ERDF) through NORTE 2020 (North Regional Operational Program 2014/2020). For the authors integrated with the CITAB Research Centre, it was further financed by the FEDER/COMPETE/POCI – Operational Competitiveness and Internationalisation Programme, under Project POCI-01-0145-FEDER-006958, and by the National Funds of FCT – Portuguese Foundation for Science and Technology, under the project UID/AGR/04033/2020. For the author integrated in the CQVR, the National Funds of FCT – Portuguese Foundation for Science and Technology, under the project UID/QUI/00616/2019, supported the research. Financial support was also provided by the FCT-Portuguese Foundation for Science and Technology (Grant: SFRH/BD/146151/2019) to António Fernandes.

REFERENCES
[1]   Trösch, W., Water treatment. *Technology Guide: Principles – Applications – Trends*, Springer Science and Business Media, pp. 394–397, 2009.
[2]   Taylor, S.D., He, Y. & Hiscock, K.M., Modelling the impacts of agricultural management practices on river water quality in Eastern England. *J. Environ. Manage.*, **180**, 2016.
[3]   Pacheco, F.A.L., Martins, L.M.O., Quininha, M., Oliveira, A.S. & Sanches Fernandes, L.F., An approach to validate groundwater contamination risk in rural mountainous catchments: The role of lateral groundwater flows. *Methods X*, 2018.
[4]   Nas, S.S. & Nas, E., Water quality modeling and dissolved oxygen balance in streams: A point source Streeter–Phelps application in the case of the Harsit stream. *Clean – Soil, Air, Water*, 2009.
[5]   Vannote, R.L., Minshall, G.W., Cummins, K.W., Sedell, J.R. & Cushing, C.E., The river continuum concept. *Can. J. Fish. Aquat. Sci.*, 1980.
[6]   Martins, L., Pereira, A., Oliveira, A., Fernandes, A., Sanches Fernandes, L.F. & Pacheco, F.A.L., An assessment of groundwater contamination risk with radon based on clustering and structural models. *Water*, **11**, p. 1107, 2019.
[7]   Sousa, J.C.G., Ribeiro, A.R., Barbosa, M.O., Ribeiro, C., Tiritan, M.E., Pereira, M.F.R. & Silva, A.M.T., Monitoring of the 17 EU watch list contaminants of emerging concern in the Ave and the Sousa Rivers. *Sci. Total Environ.*, 2019.

[8] Sanches Fernandes, L.F., Fernandes, A.C.P., Ferreira, A.R.L., Cortes, R.M.V. & Pacheco, F.A.L., A partial least squares: Path modeling analysis for the understanding of biodiversity loss in rural and urban watersheds in Portugal. *Sci. Total Environ.*, **626**, pp. 1069–1085, 2018.

[9] Gupta, S.K., *Modern Hydrology and Sustainable Water Development*, 2010.

[10] Garnier, J., Brion, N., Callens, J., Passy, P., Deligne, C., Billen, G., Servais, P. & Billen, C., Modeling historical changes in nutrient delivery and water quality of the Zenne River (1790s–2010): The role of land use, waterscape and urban wastewater management. *J. Mar. Syst.*, 2013.

[11] Hayet, C., Saida, B.A., Youssef, T. & Hédi, S., Study of biodegradability for municipal and industrial Tunisian wastewater by respirometric technique and batch reactor test. *Sustain. Environ. Res.*, **26**, pp. 55–62, 2016.

[12] Sheoran, A.S. & Sheoran, V., Heavy metal removal mechanism of acid mine drainage in wetlands: A critical review. *Miner. Eng.*, **19**, pp. 105–116, 2006.

[13] Gonçalves, E.P.R., Boaventura, R.A.R. & Mouvet, C., Sediments and aquatic mosses as pollution indicators for heavy metals in the Ave River Basin (Portugal). *Sci. Total Environ.*, 1992.

[14] Hernando, M.D., Mezcua, M., Fernández-Alba, A.R. & Barceló, D., Environmental risk assessment of pharmaceutical residues in wastewater effluents, surface waters and sediments. *Proceedings of the Talanta*, 2006.

[15] Grizzetti, B., Bouraoui, F. & De Marsily, G., Assessing nitrogen pressures on European surface water. *Global Biogeochem. Cycles*, **22**, 2008.

[16] Serrano-Grijalva, L., Sánchez-Carrillo, S., Angeler, D.G., Sánchez-Andrés, R. & Álvarez-Cobelas, M., Effects of shrimp-farm effluents on the food web structure in subtropical coastal lagoons. *J. Exp. Mar. Bio. Ecol.*, **402**, pp. 65–74, 2011.

[17] Heathwaite, A.L., Dils, R.M., Liu, S., Carvalho, L., Brazier, R.E., Pope, L., Hughes, M., Phillips, G. & May, L., A tiered risk-based approach for predicting diffuse and point source phosphorus losses in agricultural areas. *Sci. Total Environ.*, 2005.

[18] Hooda, P.S., Edwards, A.C., Anderson, H.A. & Miller, A., A review of water quality concerns in livestock farming areas. *Sci. Total Environ.*, 2000.

[19] Goldberg, V.M., Groundwater pollution by nitrates from livestock wastes. *Environ. Health Perspect.*, 1989.

[20] Huang, Z., Han, L., Zeng, L., Xiao, W. & Tian, Y., Effects of land use patterns on stream water quality: A case study of a small-scale watershed in the Three Gorges Reservoir Area. *China. Environ. Sci. Pollut. Res.*, 2016.

[21] Ferreira, A.R.L., Sanches Fernandes, L.F., Cortes, R.M.V. & Pacheco, F.A.L., Assessing anthropogenic impacts on riverine ecosystems using nested partial least squares regression. *Sci. Total Environ.*, **583**, pp. 466–477, 2017.

[22] Soares, H.M.V.M., Boaventura, R.A.R., Machado, A.A.S.C. & Esteves Da Silva, J.C.G., Sediments as monitors of heavy metal contamination in the Ave River Basin (Portugal): Multivariate analysis of data. *Environ. Pollut.*, **105**, pp. 311–323, 1999.

[23] Alves, C., Boaventura, R. & Soares, H., Evaluation of heavy metals pollution loadings in the sediments of the Ave River Basin (Portugal). *Soil Sediment Contam.*, **18**, pp. 603–618, 2009.

[24] Fonseca, A.R., Sanches Fernandes, L.F., Fontainhas-Fernandes, A., Monteiro, S.M. & Pacheco, F.A.L., From catchment to fish: Impact of anthropogenic pressures on gill histopathology. *Sci. Total Environ.*, **550**, pp. 972–986, 2016.

[25] Ribeiro, C.M.R., Maia, A.S., Ribeiro, A.R., Couto, C., Almeida, A.A., Santos, M. & Tiritan, M.E., Anthropogenic pressure in a Portuguese river: Endocrine-disrupting compounds, trace elements and nutrients. *J. Environ. Sci. Heal.: Part A Toxic/Hazardous Subst. Environ. Eng.*, 2016.

[26] Dunck, B., Lima-Fernandes, E., Cássio, F., Cunha, A., Rodrigues, L. & Pascoal, C., Responses of primary production, leaf litter decomposition and associated communities to stream eutrophication. *Environ. Pollut.*, 2015.

[27] ESRI ArcMap 10.1, *Environ. Syst. Resour. Inst.*, 2012.

[28] ESRI ArcHydro Tools for ArcGIS 10: Tutorial, 2012.

[29] De Jesus, H., Sousa, R., Oliveira, R. & Nery, F., *A Componente Geográfica do SNIRH. IV SILUSBA*, 1999.

[30] Adamczyk, J. & Tiede, D., ZonalMetrics: A python toolbox for zonal landscape structure analysis. *Comput. Geosci.*, **99**, pp. 91–99, 2017.

[31] Shi, P., Zhang, Y., Li, Z., Li, P. & Xu, G., Influence of land use and land cover patterns on seasonal water quality at multi-spatial scales. *CATENA*, **151**, pp. 182–190, 2017.

[32] Dow, C.L., Arscott, D.B. & Newbold, J.D., Relating major ions and nutrients to watershed conditions across a mixed-use, water-supply watershed. *J. North Am. Benthol. Soc.*, **25**, pp. 887–911, 2006.

[33] Yong, S.T.Y. & Chen, W., Modeling the relationship between land use and surface water quality. *J. Environ. Manage.*, 2002.

[34] Uuemaa, E., Roosaare, J. & Mander, Ü., Scale dependence of landscape metrics and their indicatory value for nutrient and organic matter losses from catchments. *Ecol. Indic.*, 2005.

[35] Lee, S.W., Hwang, S.J., Lee, S.B., Hwang, H.S. & Sung, H.C., Landscape ecological approach to the relationships of land use patterns in watersheds to water quality characteristics. *Landsc. Urban Plan.*, 2009.

[36] Yu, S., Xu, Z., Wu, W. & Zuo, D., Effect of land use types on stream water quality under seasonal variation and topographic characteristics in the Wei River Basin, China. *Ecol. Indic.*, **60**, pp. 202–212, 2016.

[37] Mahmud, Z.H. et al., Occurrence of Escherichia coli and faecal coliforms in drinking water at source and household point-of-use in Rohingya camps, Bangladesh. *Gut Pathog.*, **11**, 2019.

[38] U.S. Environmental Protection Agency, Aquatic life ambient water quality criteria for ammonia: Freshwater. *Off. Water, Sci. Technol.*, **70**, 2013.

[39] Araújo, M.F., Valério, P. & Jouanneau, J.M., Heavy metal assessment in sediments of the Ave River Basin (Portugal) by energy-dispersive x-ray fluorescence spectrometry. *X-Ray Spectrom.*, 1998.

[40] Neary, D.G., Ice, G.G. & Jackson, C.R., Linkages between forest soils and water quality and quantity. *For. Ecol. Manage.*, **258**, pp. 2269–2281, 2009.

[41] Ahearn, D.S., Sheibley, R.W., Dahlgren, R.A., Anderson, M., Johnson, J. & Tate, K.W., Land use and land cover influence on water quality in the last free-flowing river draining the western Sierra Nevada, California. *J. Hydrol.*, **313**, pp. 234–247, 2005.

[42] Pissarra, T.C.T., Valera, C.A., Costa, R.C.A., Siqueira, H.E., Martins Filho, M.V., Valle Júnior, R.F. do, Sanches Fernandes, L.F. & Pacheco, F.A.L., A regression model of stream water quality based on interactions between landscape composition and riparian buffer width in small catchments. *Water*, 2019.

STATUS OF DISCHARGED ABATTOIR EFFLUENT AND ITS EFFECTS ON THE PHYSICO-CHEMICAL CHARACTERISTICS OF OROGODO RIVER, DELTA STATE, NIGERIA

GODWIN ASIBOR, OGHENEKOHWIRORO EDJERE & CHRISTOPHER AZUBUIKE
Department of Environmental Management and Toxicology, Federal University of Petroleum Resources, Nigeria

ABSTRACT
The discharge of untreated wastewater into waterbodies results in water quality deterioration of the receiving waters. This study assesses the impact of abattoir wastewater discharge on the water quality of the Orogodo River in Nigeria. Effluent discharges and water samples were collected from the river at six points over a 6-month period. Physicochemical analyses were conducted using standard methods. The pH was within a fixed band of 5.56–8.04. The downstream biochemical oxygen demand of the receiving river water increased significantly to 75% in July and up to 192% in December. Suspended solids, chemical oxygen demand, total nitrogen and total phosphorus followed a similar trend. Dissolved solids, dissolved oxygen and nitrate also increased appreciably. The downstream levels of these parameters were higher than their corresponding upstream values, indicating that the discharge of the abattoir wastewater into the river has negatively impacted the river water. The dilution of the waste in the river water was not enough to reduce them to acceptable levels. This study demonstrates that abattoir wastewater impacts Orogodo River water negatively. The abattoir effluent did not meet the National standard for effluent discharge into the environment leading to cross pollution of the receiving water based on the parameters investigated. This, therefore, calls for the need to put an effective wastewater treatment and monitoring system in place to enforce existing legislations to curb water pollution and to safeguard both the environment and human health.
Keywords: abattoir, wastewater, physico-chemical, impact, Orogodo, pollution, effluent, parameters, discharge.

1 INTRODUCTION

An abattoir is a facility approved and registered by the controlling authority for hygienic slaughtering and inspection of animals, processing and effective preservation and storage of meat products for human consumption [1]. The abattoir industry provides domestic meat supply to over 150 million people and employment opportunities for the Nigerian teaming population [2]. Discharge wastes from abattoir can be valuable for crops as fertilizers but can becomes a major pollutant when the wastes are not properly managed [3].

According to Mittal [4] and Adeyemi-Ale [5] waste generated at abattoirs pose a serious threat to the environment because of direct discharges of wastewaters into the ecosystems which most times are not effectively treated. These wastes are high in organics and fats [6]. These result to the destruction of primary producers in the water. It is one of greatest threats to surface water quality as it causes an increase in the biochemical oxygen demand (BOD), chemical oxygen demand (COD), total solids (TS), pH, temperature, turbidity, nitrate, phosphate, etc. [7]–[15]. Biodegradable organic matter in receiving waters create high competition for oxygen within the ecosystem leading to high levels of BOD and a reduction in dissolved oxygen (DO), which is detrimental to aquatic life and affects sediments and surrounding soil.

In a typical Nigerian abattoir, the surrounding land is often marshy due to improper channelling of wastewater arising from the dressing of the slaughtered animals and washings at the lairage [3]. Most Nigerian abattoirs are situated close to surface water bodies in order to have access to cheap water supply needed for slaughtered animal processing and to provide a sink for the run-off from meat processing activities [10], [16].

In Nigeria, like in many other developing countries, discharge of untreated wastes into the environment is a major environmental issue. Compromised water quality and poor sanitary conditions of abattoirs in the livestock sector have added in no small way to the lotic water system [17].

The two studied abattoirs are located at the bank of Orogodo River in Agbor, Delta State. The Abattoir serves the entire town, and its location beside the stream has facilitated easy disposal of the wastes into the stream channel. Wastes from the abattoir are not treated before discharge into the river. This study therefore examines the implication of the continuous discharge of these untreated abattoir effluents on the river water quality. This is seen to be justified by the fact that the downstream residents very much depend on the water for most of their domestic and commercial activities. This study therefore evaluates the water quality downstream of the river, with the aim of establishing the extent to which untreated abattoir wastes would have impacted on the stream water quality and the point at which appreciable self-purification is attained.

2 MATERIALS AND METHODS

2.1 Study area

Orogodo River lies within the humid tropical zone with defined dry season (November–March) and rainy season (April–October). The rainy season is brought about by the South-West Trade Wind blowing across the Atlantic Ocean, while the dry, dusty, and often cold North-East Trade Wind blowing across the Sahara Desert dominates the dry season with a short spell of harmattan. The relative humidity of the area is high and increases from 70% in January to 80% in July. The average atmospheric temperature of the area is about 25.5°C in the rainy season and about 30°C in the dry season.

Orogodo River is located between latitudes 5° 43'N and 5° 30'N and longitudes 6° 20'E and 6° 12'E and takes its source from Mbiri village at an elevation of 150m above sea level. The river serves as a major source of water for drinking, bathing, fishing, washing, and recreation for the people of Agbor and Owa communities in Delta State, Nigeria. The Agbor and Owa communities, through which the Orogodo River traverses, are mainly peasant farmers whose products include food stuff such as yams, corn, vegetables, cassava, plantain and fruits. Agricultural activities in the area are mostly carried out along the bank of the Orogodo River. Although Agbor and Owa may not be described as industrial communities, there exist pockets of industries (paint and foam industries) as well as many educational institutions whose wastes also find their way into the river [18]. Some physico-chemical characteristics of Orogodo River have been reported. There are two major abattoirs located along the River. The sitting of abattoir is in such a way that the effluent from their operations is discharged directly into the river where any form of treatment.

2.2 Field sampling and quality assurance

Samples were collected from six locations along Orogodo River representing the upstream, midstream and downstream of the river after the effluent outfall from the abattoir between the months of June 2016 to March 2017 using the grab sampling methods. The samples represented conditions before discharge, after discharge and further down the river with emphasis on mixing of river with abattoir effluent. Details of sample collection are captured in Table 1, while GPS readings for the sampling sites were taken and used to geo-reference the location of the sampling sites.

The physical parameters analyzed include: electrical conductivity (EC), salinity, turbidity, total dissolve solid, alkalinity, temperature and pH. The chemical parameters analyzed include: nitrate, chloride, sulphate, iron, phosphate, chemical oxygen demand (COD), biological oxygen demand (BOD), and dissolved oxygen (DO). The bacteriological parameters analyzed are total coliform and faecal coliform. A total number of sixty water samples were collected along the river channel during the seven-month sampling period. Samples were collected in plastic bottles, pre-cleaned by washing with non-ionic detergents, rinsed with deionized water prior to usage. Samples for BOD and COD analysis were collected in BOD bottles and plastic bottles and covered with aluminum foil. The sample bottles were labelled according to sampling sites. All samples were preserved at 4°C and transported to Federal University of Petroleum Resources Effurun (FUPRE) Environmental Management and Toxicology (EMT) laboratory for analyses within 24 hours. The following parameters were determined in-situ: conductivity (EC), temperature, colour, odour, pH, turbidity and dissolved oxygen (DO) using the respective calibrated in-situ kits and meters.

2.3 Laboratory water analysis

The chemical oxygen demand (COD), nitrate (NO_{-3}), phosphate (PO_4^{-3}), and sulfate (SO_4^{-2}) were measured using spectrophotometer. Biochemical oxygen demand (BOD_5), total dissolved solids (TDS), total suspended solids (TSS), chloride (Cl-), bicarbonate (HCO3-), calcium, magnesium, sodium and potassium were determined using the standard methods [19], [20].

2.4 Bacteriological assessment of water and effluent samples

About 100 ml of the water sample was filtered through a filter that retains bacteria. The filtrate was then transferred to petri dishes containing MacConkey agar and incubated at 37°C for 48 hours as described by Padilla-Gasca et al. [21]. The numbers of coliform colonies formed were counted using a microscope and the values expressed as cfu/ml.

2.5 Data analysis

The data were analyzed with Microsoft Office Excel and SPSS software version 16.0.

Table 1: Sampling coordinates along Orogodo River.

Station	Location name	Coordinates		Status
		Northings	Eastings	
Station 1	Mr. Biggs	06°16'20''	006°11'16''	Upstream
Station 2	Aliumeze	06°16'20''	006°11'31''	Upstream
Station 3	Madam Jacky Slaughter	06°16'25''	006°11'31''	Discharge
Station 4	MC Commander Slaughter	06°15'25''	006°11'20''	Discharge
Station 5	Iyama-1	06°15'13''	006°11'16''	Downstream
Station 6	Iyama-2	06°13'46''	006°10'44''	Downstream

3 RESULTS AND DISCUSSION

3.1 Abattoir effluent

The colour and odour were objectionable; an indication of from the direct dumping of raw blood, fresh and decaying flesh as well as dung from the abattoir into the receiving river without any form of treatment. The untreated abattoir wastewater temperature was slightly warm, ranging from 30.5 to 33.3°C. Temperature is one of the most important factors influencing chemical and biological characteristics of water. The high temperatures recorded in the untreated abattoir wastewater aids in quick bacteria activities of the products and also assist in the breakdown of wastes into further constituencies. TSS and turbidity varied from 175–480 mg/l and 104–218 mg/l with mean concentrations of 320.4±34.8 mg/l and 202.2±22.5 mg/l. The high concentrations of TSS and turbidity are not surprising as these are products of direct discharges from the abattoirs [9]–[14], [22]. The presence of such high concentration of TSS and turbidity reduces the aesthetic value of the receiving water bodies and reduce DO of the river, while the objectionable colour and odour reduces the aesthetic and potability values of the water.

The pH varied from slightly acidic to slightly alkaline varying from 6.56 to 7.78 with an overall average of 6.74±0.54 throughout the sampling period. Wastewater from abattoir generally tends to be alkaline as a result of a high concentration of organic compounds present which is composed mainly of proteins [21], [23]. A similar result was obtained by [7]–[9], [24]. Conductivity and TDS showed same trend as they were all high in the untreated abattoir wastewater.

DO was low (<1.00 mg/l) in all the locations sampled. DO is a very important parameter for the survival of aquatic organism and is also used to evaluate the degree of freshness of a river [25]. The low value of DO is as a result of the high nutrients contents from the abattoir waste [1], [24]. The BOD was slightly high varying from 88.5 mg/l to 158.5 mg/l. The strength of domestic sewage is measured in terms of its BOD level. It determines the amount of dissolved oxygen consumed by aerobic bacteria in the decomposition of organic matter in the sewage. A typical wastewater BOD in Nigeria occurred within a range of 200–500 mg/l [26]. The COD concentration of the wastewater varied from 775–1175 mg/l. Using [27] classification the wastewater can be categorized to be of slightly high in strength. Both BOD and COD are highly related to DO as well as to each other because the BOD and COD directly affect the amount of DO in the river. DO, BOD and COD are important water quality parameters and are very essential in water quality assessment [28]. They are used to determine whether a water body is polluted or not. The higher the BOD and COD values, the higher the depletion of DO in the receiving water by organic and inorganic pollutants present in the effluents. According to Practi et al. [29], BOD, DO, Chloride, pH and Nitrate are some of the chemical parameters generally used as indices.

The nutrients parameters were observed to be slightly high in the wastewater as indicated in the overall results of nitrate, phosphate, sulphate and chlorides (Table 2). One other major concern observed from the abattoir wastewater discharge to river channel is the high presence of pathogenic bacteria, though this is not strange. Faecal coliform and total coliform bacteria are used as indicators of bacterial contamination of the rivers. The high bacteria populations recorded reflect the input of microorganisms from both the abattoir wastewater and other sources into the river as there is no toilet facilities for the operators.

Table 2: Physicochemical and bacteriological results of the Orogodo River and discharged Abattoir effluent.

Parameters	Station 1	Station 2	Station 3	Station 4	Station 5	Station 6
pH	5.73±0.43	6.59±0.25	6.72±0.37	6.76±0.03	5.64±0.38	5.34±0.66
Temp (°C)	26.72±2.61	25.26±2.4	32.5±0.82	33.3±0.74	25.26±2.35	25.5±2.0
Conductivity (μs/cm)	72.8±6.85	74.54±3.4	780±48.3	841±32.0	85.69±6.78	74.8±4.2
TDS (mg/l)	65±4.33	69.32±3.56	677±52.1	750±47.1	77.94±8.45	67.4±3.2
TSS (mg/l)	14.6±4.09	16.47±3.22	185±12.7	227.2±26.5	15.81±1.19	15.2±1.88
DO (mg/l)	2.53±0.39	3.21±0.26	0.79±0.23	0.2±0.01	3.89±0.57	4.28±0.41
BOD (mg/l)	11.8±1.89	11.27±0.79	95±4.97	112.5±4.2	7.83±2.08	6.87±1.39
COD (mg/l)	78.6±3.68	88.7±4.76	964±8.43	1045±42.2	110.2±14.2	61.1±10.4
Turbidity (N.T.U)	26.5±6.18	22.4±1.26	108±4.83	150±4.71	23.7±1.2	24.3±1.53
PO_4^{3-} (mg/l)	0.06±0.05	0.05±0.03	5.49±0.17	10.55±0.64	0.04±0.03	0.03±0.01
Cl^- (mg/l)	0.62±0.07	3.55±1.10	46.8±3.2	49.3±1.03	0.96±1.02	0.55±1.16
SO_4^{2-} (mg/l)	1.94±1.23	2.05±0.67	15.6±4.6	21.5±2.54	1.69±0.96	1.36±0.38
NO^3-N(mg/l)	0.07±0.01	0.12±0.02	8.13±0.39	10.9±0.84	0.08±0.01	0.08±0.03
HCO_3^- (mg/l)	4.26±0.97	11.15±2.37	48.08±5.68	75.6±4.22	4.81±2.64	7.61±1.67
FC (cfu/100 ml)	11.8±3.01	14.4±2.63	32500±8031	54000±79801	19.3±6.2	13.7±2.1

3.2 River upstream water characteristics

The mean summary of physico-chemical and bacteriological results obtained from the analysis of Orogodo River are presented in Table 2. It was observed that among the physical parameters the water was colourless and odourless, while the temperature was slightly low (less than 30°C) in all the locations during the sampling period.

Total suspended solids and turbidity was also observed to be low when compared to other Stations. pH was observed to be slightly acidic (5.73±0.44 and 6.59±0.25), while conductivity and TDS was low. These are characteristic of most lakes and streams of the world [30]–[32]. The nutrients load follows same pattern as observed in the conductivity and TDS as they were all low. Dissolved oxygen, BOD and COD were low indicating low oxygen demand on the water. Faecal coliform and total coliform bacteria were also recorded from this section of the river, though at a lower count compared to other Stations. The presence of faecal bacteria in this section of the river is due to open defecation practiced by some of the inhabitants of the area who do not have access to good waste facilities and organic material washed from the land into the river.

3.3 River downstream water characteristics

Downstream of the river after discharged effluent, there was observed changes in the colour and odour of the receiving river. The river water colour was greyish yellow, while the odour was still objectionable. The water temperature was slightly tepid (>25°C) less than the wastewater temperature (33.3 ± 0.78°C). The pH was slightly acidic, while conductivity and TDS showed a significant drop in concentration when compared to the effluent concentrations. Oxygen demanding parameters and nutrients followed same pattern observed for TDS and conductivity. A gradual reduction in the physico-chemical load of the water as one move downstream of the river was observed in Stations these stations. Objectionable odours and colours were no longer observed in the stations and a drastic reduction in the TSS, temperature and turbidity were also indicated from the obtained results. Relatively low values were recorded for turbidity, BOD, COD, sulphate, nitrate and phosphates compared to the upstream river water.

3.4 Comparison of obtained values to regulatory standards

Comparison of mean data set for Orogodo River before and after effluent discharge and regulatory standards are shown in Table 3. The pII of all the samples collected indicated slightly acidic environment. The pH range of this study is low comparable to pH ranges of 6.9–8.8 of previous studies on effluent from similar abattoirs in Nigeria [33], [34]. It is also slightly below most of the regulatory limits except Stations 3 and 4. pH is important to microorganisms because it affects the functioning of virtually all enzymes, hormones and proteins which control metabolism, growth and development.

Table 3: Comparison of the abattoir effluents with regulatory standards.

Parameters	Station 1	Station 2	Station 3	Station 4	Station 5	Station 6	FEPA (1991)	WHO (2004)
pH	5.73	6.586	6.72	6.76	5.641	5.34	7	6.5–8.5
Temp (°C)	26.72	25.26	32.5	33.3	25.26	25.5	<40°C	24–30
Conductivity (μs/cm)	72.8	74.54	780	841	85.693	74.783	NS	500
TDS (mg/l)	65	69.32	677	750	77.94	67.44	2	0
TSS (mg/l)	14.6	16.47	185	227	15.81	15.21	-	-
DO (mg/l)	2.53	3.21	0.795	0.2	3.89	4.28	<0.007	NS
BOD (mg/l)	11.8	11.27	95	113	7.83	6.87	50	>4.00
COD (mg/l)	78.6	88.7	964	1045	110.18	61.14	<75	10
Turbidity (NTU)	26.5	22.4	108	150	23.7	24.27	NS	40
Phosphorus (mg/l)	0.056	0.054	5.49	10.6	0.0366	0.0307	5	NS
Cl⁻ (mg/l)	0.623	3.554	46.8	49.3	0.9598	0.5466	600	NS
SO_4^{2-} (mg/l)	1.937	2.053	15.6	21.5	1.6943	1.3643	500	250
NO_3-N (mg/l)	0.074	0.116	8.13	10.9	0.0782	0.0836	20	400
HCO_3^- (mg/l)	4.26	11.15	48.08	75.6	4.814	7.608	<600	45
FC (cfu/100 ml)	11.8	14.4	32500	54000	19.3	13.7	NS	NS

It is also a major factor in all chemical reactions associated with formation, alteration and dissolution of minerals. It was observed that most of the physic-chemical parameters analysed for the stations along the River course were outside the regulatory limits except Station 6 which has close to 80% compliance with most of the regulatory limits. These results may be due to the impact of the abattoir effluent discharges. This might have been as a result of the diluting and self-purification of the river as one move downstream.

3.5 Seasonal comparison of the sampled locations

Tables 4 and 5 showed the results from the seasonal variation of the physico-chemical parameters of Orogodo River and the discharged effluent load. Effects on the aesthetic of the effluents were the affected by the season as indicated by the objectionable odour and colour of the effluent discharges on the River. Other physical parameters like temperature, TSS and turbidity were higher in the dry season compared to the rainy season. This may be as a result of the diluting effect of rain and drainage water from the river catchment basin.

This also contributed to the observed higher conductivity and TDS in the dry season results. The diluting effects of the rain were also reflected in the concentrations of oxygen demanding parameters, nutrients and abundance level of faecal coliforms in the sampling stations. Generally, it was observed the physic-chemical load of the river and effluent was higher in the dry season compared to the rainy season.

Table 4: Orogodo River physico-chemical and bacteriological results (rainy season).

Parameter	Station 1	Station 2	Station 3	Station 4	Station 5	Station 6
pH	5.43±0.9	6.66±0.02	6.75±0.0	6.75±0.0	5.81±0.45	5.71±0.66
Temp (°C)	24.80±0.0	25.80±2.53	32.50±0.0	33.40±0.0	25.72±2.35	26.20±2.50
Conductivity (µs/cm)	68.40±0.0	73.13±3.92	750.00±0.0	825.0±0.0	80.99±4.73	72.07±4.05
TDS (mg/l)	62.40±0.0	69.60±4.45	670.00±0.0	750.0±0.0	73.28±4.58	66.08±3.80
TSS (mg/l)	12.20±0.0	17.70±3.92	180.00±0.0	220.0±0.0	16.02±1.51	16.02±2.24
DO (mg/l)	2.30±0.0	3.28±0.10	0.88±0.0	0.20±0.0	3.66±0.67	4.06±0.46
BOD (mg/l)	10.50±0.0	11.40±0.80	95.50±0.0	110.5±0.0	8.66±2.50	7.24±1.79
COD (mg/l)	77.90±0.0	90.50±4.27	960.00±0.0	1025±0.0	102.66±15.8	64.78±13.0
Turbidity (N.T.U)	23.80±0.0	22.20±0.49	105.00±0.0	150.0±0.0	23.40±1.50	23.44±1.68
T.Phosphorus (mg/l)	0.02±0.0	0.05±0.02	5.50±0.0	10.30±0.0	0.05±0.04	0.04±0.01
Cl^- (mg/l)	0.58±0.0	3.89±0.57	45.50±0.0	49.10±0.0	1.30±1.28	0.96±1.45
SO_4^{2-} (mg/l)	1.14±0.0	2.31±0.80	14.00±0.0	20.50±0.0	2.05±1.11	1.59±0.40
NO^3-N(mg/l)	0.07±0.0	0.13±0.01	8.15±0.0	10.50±0.0	0.08±0.02	0.10±0.40
HCO_3^- (mg/l)	3.66±0.0	11.60±0.80	47.08±0.0	77.60±0.0	5.77±3.25	7.90±2.20
FC (cfu/100 ml)	10.00±0.0	13.80±3.43	32000±4000	25400±3200	16.20±5.98	13.40±1.96

Table 5: Orogodo River physico-chemical and bacteriological results (dry season).

Parameter	Station 1	Station 2	Station 3	Station 4	Station 5	Station 6
pH	6.47±0.85	6.87±0.69	7.24±1.09	6.81±0.09	5.47±0.0	5.81±1.41
Temp (^{0}C)	31.10±4.84	26.8±4.05	33.72±2.41	34.30±2.15	27.04±4.4	24.80±0.0
Conductivity (μs/cm)	84.76±14.8	77.32±2.7	864.77±107	903.04±90.5	94.87±8.8	77.50±0.0
TDS (mg/l)	72.62±9.87	70.92±3.69	761.33±151	820.71±139	91.5±17.6	68.80±0.0
TSS (mg/l)	21.82±9.47	15.82±1.14	207.32±34.1	272.44±74.8	16.1±0.88	14.40±0.0
DO (mg/l)	3.22±0.90	3.50±0.70	1.02±0.62	0.21±0.03	4.30±0.35	4.50±0.0
BOD (mg/l)	15.05±3.83	11.90±1.49	101.9±214.5	120±10.7	7.45±0.88	6.50±0.0
COD (mg/l)	84.71±10.6	91.37±8.79	978.95±21.5	1119.8±107	118±0.88	57.50±0.0
Turbidity (N.T.U)	37.43±16.2	24.39±3.52	116.48±10.7	157.1±13.9	24.5±0.88	25.10±0.0
T.Phosphorus (mg/l)	0.14±0.09	0.09±0.07	5.73±0.49	11.67±1.70	0.03±0.02	0.02±0.0
Cl⁻ (mg/l)	0.75±0.16	4.63±2.79	52.44±8.53	51.02±2.98	0.71±0.18	0.14±0.0
SO_4^{2-} (mg/l)	4.08±2.64	1.99±0.38	23.62±12.6	25.96±6.81	1.78±0.88	1.14±0.0
NO^3-N(mg/l)	0.08±0.01	0.12±0.03	8.69±1.13	12.40±2.15	0.08±0.01	0.07±0.0
HCO_3^- (mg/l)	5.96±2.15	14.06±6.63	57.45±16.4	79.08±10.7	4.31±0.88	7.32±0.0
FC (cfu/100 ml)	17.1±6.90	15.00±0.0	44157±21940	193373±217823	26.5±8.53	16.24±4.4

4 CONCLUSION AND RECOMMENDATIONS

The effects of abattoir wastewater discharge into Orogodo River on its water quality were assessed through water quality monitoring. Findings from the study indicate that livestock processing and activities around the Abattoir have impacted the river water quality. Concentrations of oxygen demanding parameters, nutrients and faecal coliforms were in excess of normal levels for the river water. The downstream levels of these parameters were higher than their corresponding upstream values, indicating that the discharge of the abattoir wastewater into the river has negatively impacted water. The dilution of the high-strength abattoir wastewater in the river water was not enough to reduce most of the physico-chemical parameters and bacterial load to acceptable levels. Although there is a potential that an improvement of the water quality may be observed further downstream due to self-purification and further dilution effects, the high levels of these parameters is a worrying issue to the riparian users around the abattoir.

This water quality data and pollution source information will be useful in identifying water quality intervention measures. The findings can also be used as basis for strengthening existing legislation.

REFERENCES

[1] Alonge, D.O., *Textbook of Meat Hygiene in the Tropics,* Farmcoe Press: Ibadan, pp. 98–105, 1991.
[2] Nafaranda, W.D., Yaji, A. & Icubkomawa, H.I., Impact of abattoir wastes on aquatic life. *Global Journal of Pure and Applied Sciences,* **12**(1), pp. 31–33, 2005.
[3] Asibor, G., Edjere, O. & Ofejiro, P., Physico-chemical and bacteriological assessment of abattoir effluents and its effects on Agbarho River, Delta State Nigeria. *Nigerian Journal of Applied Science,* **35**, pp. 265–276, 2017.
[4] Mittal, G.S., Treatment of wastewater from abattoir before land-application – a review. *Bioresource Technology,* **97**, pp. 1119–1135, 2006.

[5]   Adeyemi-Ale, O.A., Impact of abattoir effluent on the physico-chemical parameters of Gbagi Stream (Odo-Eran), Ibadan, Nigeria. *Ilorin Journal of Science*, **1**(1), pp. 100–109, 2014.

[6]   Raymond, C.L., *Pollution Control for Agriculture*, Academic Press Inc.: New York, 1977.

[7]   Cooper, R.N., Hoodle, J.R. & Russel, J.M., Characteristics and treatment of slaughterhouse effluent in New Zealand. *Prog Water Technology*, **11**, pp 55–68, 1979.

[8]   Quinn, J.M. & McFarlane, P.N., Effects of slaughterhouse and dairy factory wastewaters on epilithon: A comparison in laboratory streams. *Water Research*, **23**(10), pp. 1267–1273, 1989.

[9]   Sangodoyin, A.Y. & Agbawhe O.M., Environmental study on surface and groundwater pollutants from abattoir effluents. *Bioresource Technology*, **41**(2), pp. 193–200, 1992.

[10]   Omole, D.O. & Longe, E.O., An assessment of the impact of abattoir effluents on River Illo, Ota, Nigeria. *Journal of Environmental Science and Technology*, **1**(2), pp. 56–54, 2008.

[11]   Akpor, O.B. & Muchie, M., Environmental and public health implications of wastewater quality. *African Journal of Biotechnology*, **10**(13), pp. 2379–2387, 2011.

[12]   Ezeoha, S.L. & Ugwuishiwu, B.O., Status of Abattoir Wastes Research in Nigeria. *Nigerian Journal of Technology*. 30(2), pp. 143–148, 2011.

[13]   Adelegan, J.A., Environmental policy and slaughterhouse waste in Nigeria. *Proceedings of the 28th WEDC Conference*, Calcutta, India, pp. 3–6, 2002.

[14]   Ojo, J.O., Environmental impact assessment of effluents from Oko-Oba municipal abattoir at Agege, Lagos State, Nigeria. *Global Advanced Research Journal of Agricultural Science*, **3**(10), pp. 317–320, 2014.

[15]   Ogbonna, D.N. & Ideriah, T.J.K., Effect of abattoir wastewater on physico-chemical characteristics of soil and sediment in southern, Nigeria. *Journal of Scientific Research and Reports*, **3**(12), pp. 1612–1632, 2014.

[16]   Omole, D.O. & Ogbiye, A.S., An evaluation of slaughterhouse wastes in south-west Nigeria. *American Journal of Environmental Protection*, **2**(3), pp. 85–89, 2013.

[17]   Adeyemo, O.K., Ayodeji, I.O. & Aiki-Raji, C.O., The water quality and sanitary conditions in a major abattoir (Bodija) in Ibadan, Nigeria. *African Journal of Biomedical Research*, **5**, pp. 51–55, 2002.

[18]   Edjere, O. & Iyekowa, O., Assessment of the levels of polychlorinated biphenyls (PCBs) in Orogodo River sediments Agbor, Delta State, Nigeria. *Ovidius University Annals of Chemistry*, **28**(2), pp. 25–29, 2017. https://doi.org/10.1515/auoc-2017-0005.

[19]   Ademoroti, C.M.O., *Standard Methods for Water and Effluents Analysis*, Foludex Press Ltd.: Ibadan, pp. 58–76, 1996.

[20]   APHA, AWWA & WEF, *Standard Methods for the Examination of Water and Wastewater*. 20th ed., American Public Health Association, American Water Works Association and Water Environment Federation: Washington DC., 1999.

[21]   Padilla-Gasca, E., Lopez-Lopez, A. & Gallardo-Vaidez, J., Evaluation of stability factor in the anaerobic treatment of slaughterhouse wastewater. *Journal of Bioremediation and Biodegradation*, **2**, pp. 114–124, 2011.

[22]   Kwadzah, T.K. & Iorhemen, O.T., Assessment of the impact of abattoir effluent on the water quality of River Kaduna, Nigeria. *World Journal of Environmental Engineering*, **3**(3), pp 87–94, 2015.

[23] Ogbomida, E.T., Kubeyinje, B. & Ezemonye, L.I., Evaluation of bacterial profile and biodegradation potential of abattoir wastewater. *African Journal of Environmental Science and Technology,* **10**(2), pp. 50–57, 2016.

[24] Mulu, A., Ayenew, T. & Berhe, S., Impact of Slaughterhouses Effluent on Water Quality of Modjo and Akaki River in Central Ethiopia. *International Journal of Science and Research,* **4**(3), pp. 899–907, 2013.

[25] Kolawole, O.M., Ajayi, K.T., Olayemi, A.B. & Okoh, A.I., Assessment of water quality in Asa River (Nigeria) and its indigenous *Clarias gariepinus* fish. *International Journal of Environmental Research and Public Health,* **8**(11), pp. 4332–4352, 2011.

[26] Ogedengbe, O., Technologies for Industrial waste Management. Paper for a short certificate course workshop on Environmentally Friendly Technologies for Management of Waste, *Institute of Ecology and Environmental Studies and the UNICECS,* O.A.U. Ile-Ife, Nigeria, 1990.

[27] Mara, D.D., Wastewater treatment in hot climates. *Water, Wastes and Health in Hot Climates,* eds R. Feacham, M. McGarry & D.D. Mara, John Wiley and Sons: New York, 1977.

[28] Chapman, D., *Water Quality Assessments: A Guide to the Use of Biota, Sediment and Water in Environmental Monitoring,* 2nd ed., E and FN Spon: London, pp. 626, 1996.

[29] Practi, L., Pavenello, R. & Pasarin, P., Assessment of surface water quality by a single index of pollution. *Water Research,* **5**, pp. 741–751, 1971.

[30] Akinbuwa, O., The Rotifera fauna and physico-chemical conditions of Erinle lake and its major inflows at Ede, Osun State, Nigeria. PhD thesis, Obafemi Awolowo University, 1999.

[31] Akande, A.O. & Awotoye, O., Some aspects of the ecology of Isiula Lake – a natural aquatic ecosystem in Ado-Ekiti, Nigeria. *Nigerian Journal of Botany,* **3**, pp. 221–230, 1990.

[32] Ekpeyong, E. & Adeniyi, I.F., Physical and chemical factors and net phytoplankton distribution in some tropical fishponds. *Tropical Freshwater Biology,* **5**, pp. 43–53, 1996.

[33] Osibanjo, O. & Adie, G.U., The impact of effluent from Bodija abattoir on the physico-chemical parameters of Oshunkeye stream in Ibadan City, Nigeria. *African Journal of Biotechnology,* **6**(15), pp. 1806–1811, 2007.

[34] Raheem, N.K. & Morenikeji, O.A., Impact of abattoir effluents on surface waters of the Alamuyo stream in Ibadan. *Journal of Applied Sciences and Environmental Management,* **12**(1), pp. 73–77, 2018.

SECTION 2
ASSESSING, MONITORING, MODELLING AND FORECASTING

SENTINEL-2 ANALYSIS OF FLOODED AREAS: APPLIED CASE STUDY – LA SAFOR WETLAND, SPAIN

JESÚS PENA-REGUEIRO[1], MARIA-TERESA SEBASTIÁ-FRASQUET[1],
JESÚS A. AGUILAR-MALDONADO[1], JAVIER ESTORNELL[2],
JOSÉ-ANDRÉS SANCHIS-BLAY[1], SERGIO MORELL-MONZÓ[1] & VICENT ALTUR-GRAU[1]
[1]Instituto de Investigación para la Gestión Integrada de Zonas Costeras, Universitat Politècnica de València, Spain
[2]Grupo de Cartografía GeoAmbiental y Teledetección, Universitat Politècnica de València, Spain

ABSTRACT

La Safor wetland is a representative coastal wetland in the Valencia Region (eastern Spain, Mediterranean Sea). This wetland is recognized at an international level as a Special Protection Area (SPAs) for birds and a Site of Community Importance (SCIs) (Habitats Directive, European Council Directive). The wetland is located on a detrital plain aquifer which in turn is fed by a karstic aquifer in the near limestone reliefs. The flooded surface is variable and depends on pluviometry among other factors. The objective of this study is to analyse the effects of the flooded surface on land uses by remote sensing and Airborne LiDAR data. Sentinel-2A images processed at level 1C were obtained from Copernicus. LiDAR data was used to detect the most vulnerable areas affected by floods. In the results, we analysed the impact of the maximum flooded surface on land uses. We propose several corrective actions on the drainage net based on our analysis. This methodology can be applied to other wetland areas of similar characteristics. The advantage is the high spatial resolution which makes the methodology suitable for small sized wetlands.
Keywords: *flooding, wetland, restoration, LiDAR, remote sensing.*

1 INTRODUCTION

Coastal wetlands are important ecosystems that are endangered because of intensive human pressure in these areas [1]. Wetlands provide a wide range of ecosystem services, particularly they are important providers of all water-related ecosystem services, among them their capacity to maintain and improve water quality, flood control, groundwater replenishment, shoreline stabilisation and storm protection and climate change mitigation and adaptation [2]–[4]. Wetlands play an important role in maintaining local water quality. They can act as filtering systems, removing sediment, nutrients (mainly nitrogen and phosphorus) and pollutants (such as pesticides from agricultural runoff) from water. Their holding capacity helps control floods and prevents water logging of crops and adjacent urban areas [3], [5]. Preserving and restoring them can often provide the level of flood control more affordable than dredge operations and levees [5]. In fact, there are studies that calculate the benefits of economical activities in wetlands such as intensive agriculture and shrimp farms, and they are between 60% and 75% lower – in the long term – than the benefits from wetland conservation and sustainable use [2], [6]. Nowadays, projects to restore their functions are being designed and implemented.

Remotely sensed data can provide spatial maps of water bodies, with different accuracy depending on the sensor, that can be used to increase the knowledge of these areas [7]–[10]. The analysis of remote data can be based on supervised classification and the definition of water indices and their subsequent classification using thresholds [9], [10]. From 2015 onward, Sentinel-2A/B images are available (ESA), with high temporal resolution and bands of 10 m that allow to extract small-sized water bodies [7]. LIDAR (Light Detection and Ranging or Laser Imaging Detection and Ranging) data can be used to obtain synoptic digital elevation models (DEMs). The information on wetland elevation obtained from LIDAR data is very accurate and can be used for different purposes such as habitat mapping and flood inundation mapping [11]. The accuracy of the LIDAR-derived DEM can range from 0.03 to 0.25 m depending on the vegetation cover classes [11].

The objective of this study is to analyse the effects of the flooded surface on land uses by remote sensing and Airborne LiDAR data on a Mediterranean coastal wetland "La Safor wetland". Currently, the municipalities of this wetland are competing for European funds for projects to recover hydraulic infrastructures in protected areas. So, a complete diagnosis of the areas most vulnerable to flooding is necessary.

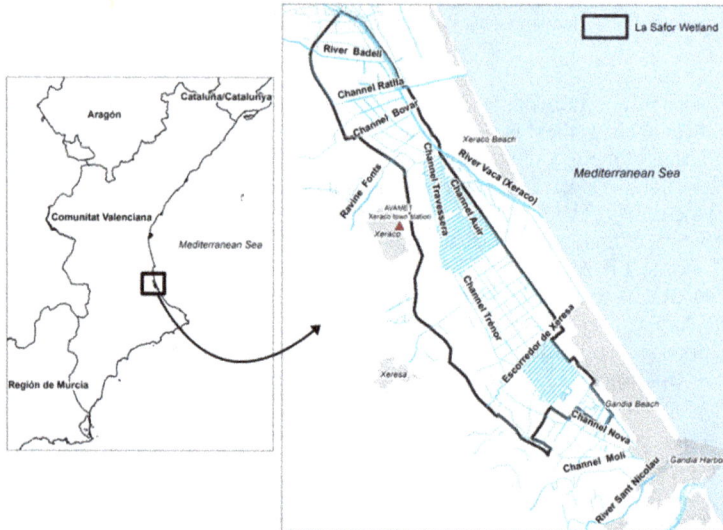

Figure 1: Study area location. The red triangle is the Xeraco town meteorological station.

2 MATERIALS AND METHODS

2.1 Study area

La Safor wetland is in south Valencia, Spain and extends to four municipalities (Gandia, Xeresa, Xeraco and Tavernes de la Valldigna) (Figs 1 and 2). It is included in both the Valencian Wetlands Inventory, and in the Spanish Wetlands, and at international level, is a Special Protection Areas (SPAs) for birds and Site of Community Importance (SCI). This coastal wetland is above a detrital aquifer, and its main freshwater input comes from groundwater discharge of the adjacent karstic aquifers and is separated from the sea by a sand bar [1]. The flooded surface is variable and depends on pluviometry among other factors [7]. La Safor wetland has suffered intense agricultural transformations mainly from the seventies onward [1]. Nowadays, the main land uses are citrus and horticultural crops according to the Corine Land Cover obtained from the Valencian Cartography Institute (Fig. 2).

Two watercourses are the main drainage of the wetland, San Nicolás and Vaca Rivers, to the north and south respectively (Fig. 1). Both rivers drain the wetland and flows into the Mediterranean Sea. San Nicolás flows into the Gandia Harbour and Vaca River flows directly into the sea in Xeraco. There is a built grid of channels that served for irrigation purposes before it has been substituted by drip irrigation. These channels also contributed to wetland drainage as they were connected either to the above described rivers or directly to the sea. However, two main factors caused the deterioration and even partial disappearance of these channels. The change in the irrigation system caused an important reduction in maintenance tasks by farmers. The urban pressure, especially in Gandia municipality, caused for example the disappearance of Escorredor de Xeresa in its final course to the sea (Fig. 3). In 1956, this channel outflow in behind the dunes to a channel parallel to the sea. Partially due to this abandonment of hydraulic infrastructures there exists flooding problems in the agricultural and urban areas.

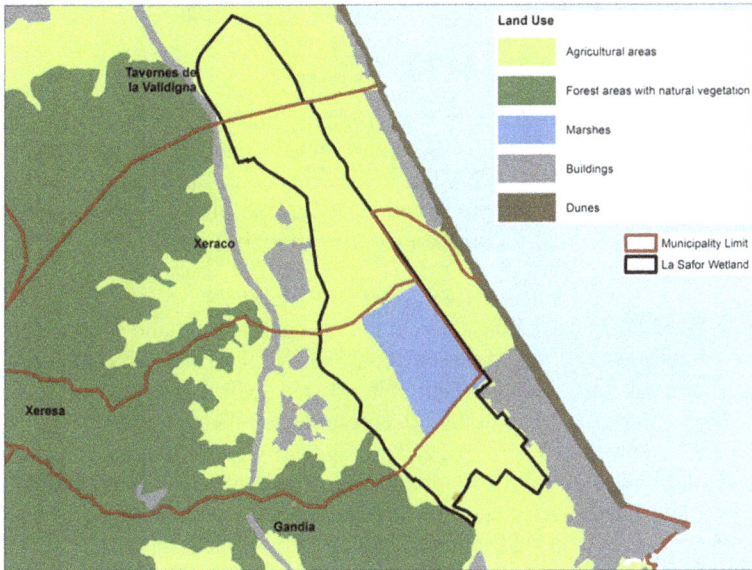

Figure 2: Land uses and municipality limits in the La Safor wetland.

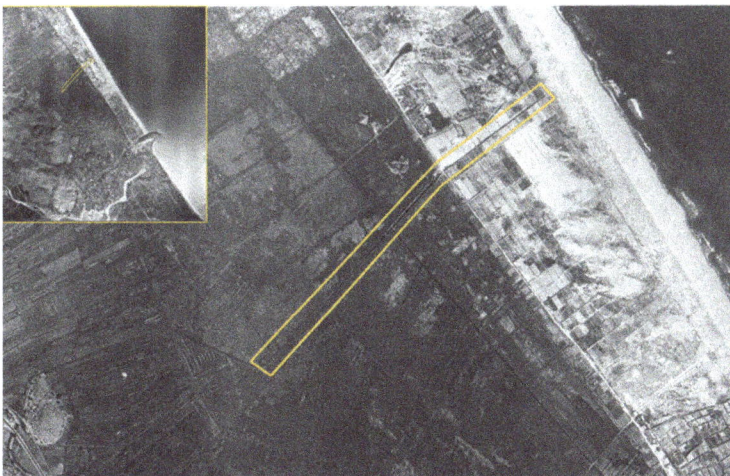

Figure 3:   Aerial image, 1956. Marked in yellow is the channel "Escorredor de Xeresa".

2.2 Image processing

The official cartography of protected areas (Valencian Wetland Inventory) was used to delimitate wetland (Figs 1 and 2). The LIDAR-derived DEM was calculated for 2 m pixel, the orthometric heights information was obtained from LiDAR sensor with a density of 0.5 points/m^2 (PNOA 2007 CC BY 4.0 www.scne.es).Sentinel-2A images from the Multispectral Instrument (MSI) processed at level 1C were obtained from Copernicus (https://scihub.copernicus.eu/dhus/#/home). The atmospheric correction was done with Sen2Cor tool (version 02.05.05) using SNAP software (ESA, version 6.0.0). The NDWI index was calculated according to Pena-Regueiro et al. [7] see eqn (1):

$$NDWI = \frac{B03 - B08}{B03 + B08}. \tag{1}$$

B03 and B08 are Sentinel-2 bands with 10 m spatial resolution. For each date, we delimited the water polygons using the -0.30, threshold defined by Pena-Regueiro et al. [7]. The acquisition date of the Sentinel 2A image was chosen to show the higher extent of the flooded area. This depends on the precipitation regime, so pluviometry data was obtained from the closest meteorological station, Xeraco town station (Fig. 1) [12].

3 RESULTS AND DISCUSSION

The LIDAR-derived DEM can be observed in Fig. 4. La Safor wetland is in a depressed area between karstic relieves and the Mediterranean Sea. Most of the protected area is below 2.5 m.a.s.l. In the north part, there is an agricultural area dedicated to horticultural crops that is below sea level (between -1.5–0 m.a.s.l.), represented by the white area in Fig. 4(b).

From the 1970s onward, a new production system called "bancs" was developed in La Safor wetland [1] that raised the height of agricultural plots with the materials dredged from adjoining plots which became ponds. Also, mining licenses for peat were granted when excavating up to one and a half meters below ground level. In the 1980s and 1990s, the agricultural transformation projects continued to create large citrus farms [1]. This transformation can be appreciated in Fig. 4, plots between 1.5–2.5 m adjoining plots between 0.5–1 m (dark blue-pink areas), and plots between 3–5 m adjoining plots between 1.5–2.5 m (green-dark blue areas). The most depressed plots correspond to the excavated ponds, and the highest to the "bancs".

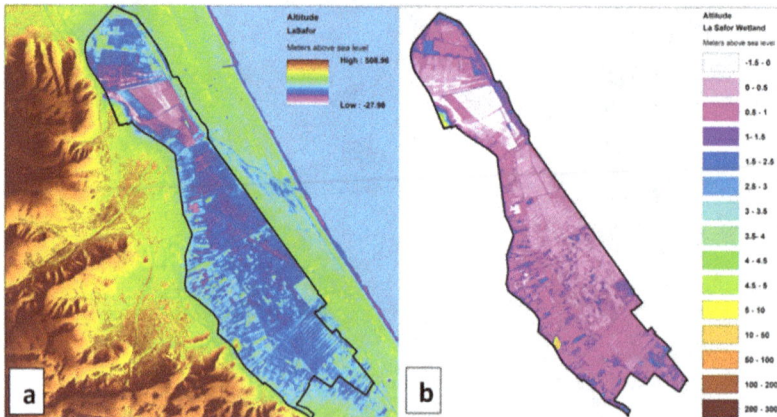

Figure 4: LIDAR-derived digital elevation model.

In Fig. 5, we can observe the results of mapping water bodies by the NDWI methodology [7]. The Sentinel-2A image was captured on January 16, 2017. The results show a total flooded surface of 1.46 km². The image data was selected for two reasons. The average annual precipitation for the Xeraco town meteorological station is 709.2 mm, and the two months before the image data there was an accumulated precipitation of 326.7 mm. Then, the image was captured after a wet period with maximum precipitation in 24 hours of 50.6 mm on November 28, 2016, 71.8 mm on December 5, 2016 and two consecutive days with 46.1 mm on December 18 and 19, 2016. Usually, dense vegetation prevents remote sensing indices to detect water, and we can only observe free vegetation water bodies. But, on November 9, 2016, there was a fire in the wetland that burned 90 Ha (Fig. 6). The disappearance of dense marsh vegetation made possible to appreciate the water layer after the rainy period. The main water polygon (1 km² flooded area) is in an area called "Les Galerasses" which is where marsh is better conserved (Fig. 2). We can also appreciate, north and south of this bigger water polygon, longitudinal polygons that correspond to the above described ponds. In the most depressed area, white polygon in Fig. 4, we do not appreciate relevant flooding. This can be explained because in these plots there is a big farm of horticultural crops, and they have four water pumps to discharge water to the Vaca river.

Local authorities (Gandia, Xeresa, Xeraco and Tavernes councils) usually need to deal with conflicting interests among different users, i.e. owners, farmers, irrigators, hunters, fishermen, residents, conservationists, and companies. Farmers are grouped in irrigation communities, such as Gandia's Irrigation Community. For these users, it is very important to avoid crops flooding, because this can cause root asphyxia and loss of productivity [13]. Then, they control several water pumps in the wetland (inside and outside of the protected area), and they turn them on after important rain event to discharge water through the irrigation channels to the sea. This evacuating capacity is not homogenous in all the wetland, Gandia has more facilities, while Xeraco and especially Xeresa have more difficulties due to the abandonment or disappearance of hydraulic infrastructures (i.e. Escorredor de Xeresa).

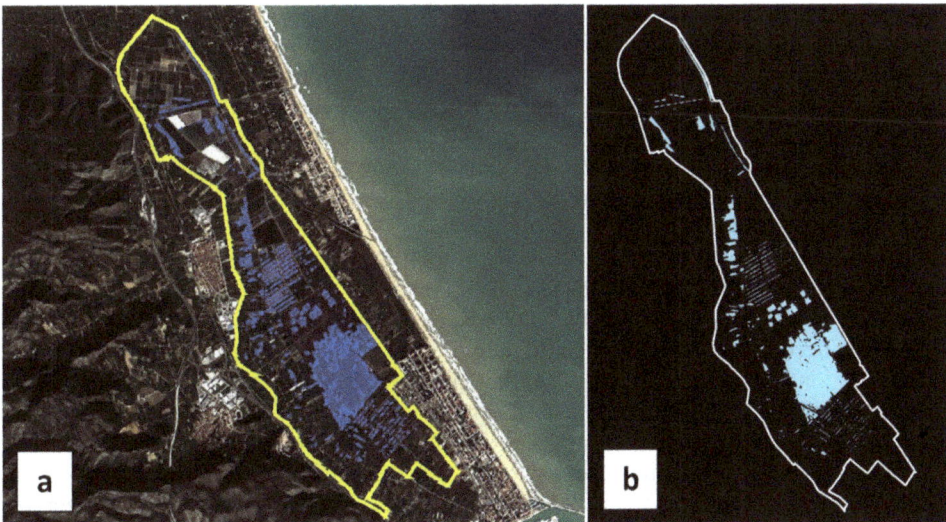

Figure 5: La Safor wetland: (a) Natural colour image and NDWI mask, January 16, 2017; and (b) NDWI mask with higher contrast, January 16, 2017.

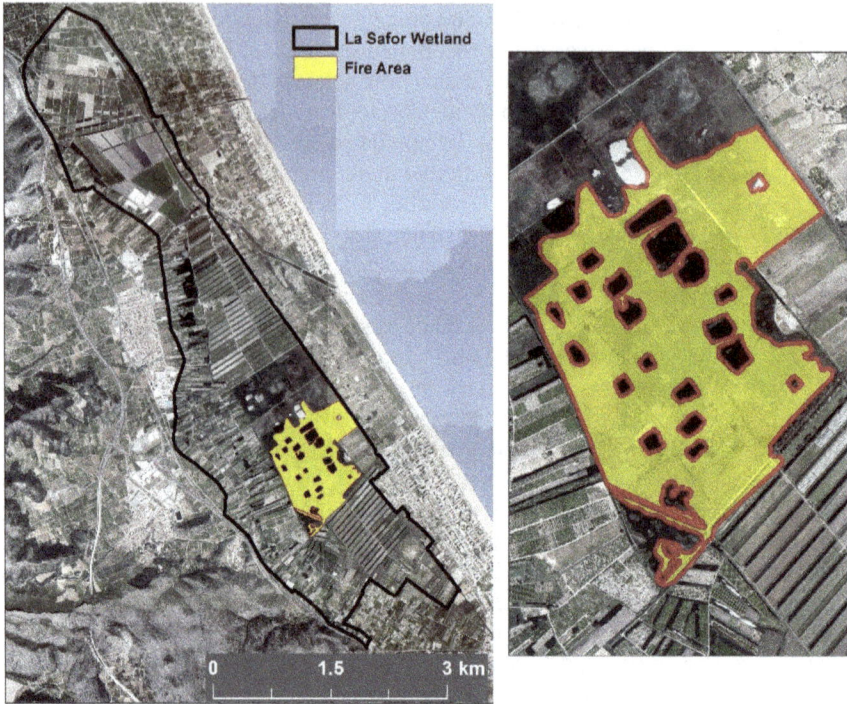

Figure 6: In yellow: fire area burned on November 9, 2016 the in La Safor wetland.

In 2014, 91.9% of La Safor Wetland was private land, divided in small properties and big farms. There is consensus that private land ownership complicates management of protected areas and limits the scope of action of the various administrations [1]. One solution to this increasing problem is the purchase of land by the administration [14]. In 2018, the Territorial Action Plan for the Green Infrastructure of the Coast of the Valencian Community was passed by Decree 58/2018, dated May 4. Thanks to the approval of this plan, the public administration has been able to buy several plots on the north limit of the Gandia beach urban area. This action makes possible the recovery of the "Escorredor de Xeresa" old watercourse through this current public land.

The corrective measures that are included in small projects competing for different fund sources, such as European funds, are based on this analysis of land uses, the LIDAR-derived DEM and the map of flooded areas by NDWI index. Corrective measure must also have into account the need of considering all conflicting uses and ecosystem services provided by the wetland for integral management. The first measure is the restoration of the Channel Travessera (Fig. 1). Currently, this is the main discharge of the wetland area in the municipalities of Xeresa and Xeraco. The restoration will be executed in different phases. The main objective is elevating the separation between the marsh and the crop area and building discharge connections between them. So, in case of important rain event the agricultural area can drain to the marsh area, and to the ponds. This measure will fulfil different purposes, it will reduce the flooding problems of the agricultural area, it will increase the flooded area in the marsh, and it will allow a higher recharge of the aquifer avoiding direct discharge to the sea.

4 CONCLUSION

The remote sensing and Airborne LiDAR data analysis allowed to identify and map the flooded surface. For successful wetland management, it is key having into account all conflicting uses and ecosystem services provided by the wetland. In wetlands coexist uses with different needs such as agricultural and environmental use, and managers need to reach an equilibrium. Agriculture is consolidated in the La Safor wetland after long time, and it has implemented some measure to avoid productivity loss that are not the best environmentally (i.e. water pumps draining the fields to the sea). After our analysis, the corrective measures proposed a search to increase the flooded marsh area and protect the fields, allowing a higher recharge of the aquifer.

ACKNOWLEDGEMENTS
The authors want to thank the Xeresa Council and Xeraco Council for sharing information.

REFERENCES
[1]   Sebastiá-Frasquet, M.-T., Altur, V. & Sanchis, J.-A., Wetland planning: Current problems and environmental management proposals at supra-municipal scale (Spanish Mediterranean coast). *Water,* **6**, pp. 620–641, 2014.
[2]   Convention on Wetlands of International Importance (Ramsar Convention). www.ramsar.org. Accessed on: 13 Jun. 2020.
[3]   Millennium Ecosystem Assessment, *Ecosystems and Human Well-Being: Synthesis,* Island Press: Washington, DC, 2005.
[4]   Mitsch, W.J., Bernal, B. & Hernandez, M.E., Ecosystem services of wetlands. *International Journal of Biodiversity Science, Ecosystem Services & Management,* **11**(1), pp. 1–4, 2015.
[5]   Why are Wetlands Important? U.S. Environmental Protection Agency, Wetlands Protection and Restoration. https://www.epa.gov/wetlands/why-are-wetlands-important. Accessed on: 13 Jun. 2020.
[6]   Balmford, A. et al., Economic reasons for conserving wild nature. *Science,* **297**(5583), pp. 950–953, 2002.
[7]   Pena-Regueiro, J., Sebastiá-Frasquet, M.-T., Estornell, J. & Aguilar-Maldonado, J.A., Sentinel-2 application to the surface characterization of small water bodies in wetlands. *Water,* **12**, pp. 1487, 2020.
[8]   Huang, C., Peng, Y., Lang, M., Yeo, I.-Y. & McCarty, G., Wetland inundation mapping and change monitoring using Landsat and airborne LiDAR data. *Remote Sensing of Environment,* **141**, pp. 231–242, 2014.
[9]   Zhou, Y. et al., Open surface water mapping algorithms: A comparison of water-related spectral indices and sensors. *Water,* **9**, pp. 256, 2017.
[10] Tian, S., Zhang, X., Tian, J. & Sun, Q., Random forest classification of wetland landcovers from multi-sensor data in the arid region of Xinjiang, China. *Remote Sensing,* **8**, pp. 954, 2016.
[11] Hladik, C. & Alber, M., Accuracy assessment and correction of a LIDAR-derived salt marsh digital elevation model. *Remote Sensing of Environment,* **121**, pp. 224–235, 2012.
[12] AVAMET. Valencian Association of Meteorology. https://www.avamet.org/. Accessed on: 13 Jun. 2020.

[13] Sebastiá, M.-T., Rodilla, M., Sanchis, J.-A., Altur, V., Gadea, I. & Falco, S., Influence of nutrient inputs from a wetland dominated by agriculture on the phytoplankton community in a shallow harbour at the Spanish Mediterranean coast. *Agriculture, Ecosystems & Environment*, **152**, pp. 10–20, 2012.

[14] Maltby, E., Acreman, M., Blackwell, M.S.A., Everard, M. & Morris, J., The challenges and implications of linking wetland science to policy in agricultural landscapes— Experience from the UK National Ecosystem Assessment. *Ecological Engineering*, **56**, pp. 121–133, 2013.

BALANCING INNOVATION AND VULNERABILITY: WATER SECURITY IN AN AGE OF CYBER-WARFARE

KATRINA PETERSEN & PETER WIELTSCHNIG
Applied Innovation & Research Team, Trilateral Research, Ireland

ABSTRACT

Through technological and organisational innovations, water services are increasingly finding new pathways to reduce water insecurity. For instance, the real-time detection of water pollution through the deployment of novel sensors can mitigate contamination problems before they become widespread. While these tools open up new forms of water security, they simultaneously create new forms of vulnerability as the connectivity and digitalisation of these infrastructures create remote pathways to control water system behaviours. The cyber-warfare capabilities of state and non-state actors are becoming increasingly sophisticated, and attacks have successfully infiltrated water systems with worrying potential. Indeed, in 2018, the US Department of Homeland Security and Federal Bureau of Investigations highlighted the threat of cyber-attacks from hostile countries on water systems, demonstrating the very real nature of these threats. This paper assesses the vulnerabilities arising from increasing interconnectivity and digitalisation in water infrastructures, paying particular attention to demographics at risk of insecurity. The paper starts by reviewing cyber-warfare practices relating to infrastructure, including their increasing frequency and sophistication. This is overlaid with a current demographic understanding of water insecurity and potential vulnerabilities to cyber-attack to identify what intersectionalities appear as new threats emerge. The paper then explores the necessary structure and value of ethical impact assessments in the design of innovative technology and practices in the water sector. In order to foster sensitivity to vulnerabilities and create avenues for incorporating scalable preventative and mitigating measures into design, a practical framework (in the form of a list of questions) is outlined. This paper finds that our understanding of water insecurity must adapt to the challenges posed by cyber-attacks. Sensitivity to the existence of these threats must be fostered and a practical framework developed to attune stakeholders to cyber-threats and assist those engaged with new technologies in the water sector.
Keywords: water security, cyber-attacks, cyber-security, intersectionality, ethics, environmental justice, vulnerability, internet of things, impact assessments, water pollution.

1 INTRODUCTION

Water security is commonly framed in terms of securing quality and quantity, and whether water is accessible, sustainable, and drinkable. For example, in 2013, the UN defined water security as "the capacity of a population to safeguard sustainable access to adequate quantities of acceptable quality water for sustaining livelihoods, human well-being, and socio-economic development" [1]. A lack of fresh water can affect the affordability of drinking water, and have a detrimental effect regarding public health, food security, human security, and political unrest [2]. But what is at stake with water insecurity and what actions need to be taken to ensure these water security goals are met can vary greatly depending on the region of focus. Wealthy nations, like those in Europe, face different forms of water insecurity than developing countries. As a result, the European Water Framework Directive, the Drinking Water Directive have re-articulated water security in contexts where access to running water is plentiful. They shift the focus from getting water to people in the first place, to maintaining the ecological and chemical quality of water and the work necessary ensure this public good. They establish definitions for sustainable use of that water, including the potential for water scarcity from shrinking water tables, increased demand from growing populations and agricultural needs (expected to be a problem across 30% of EU Member

WIT Transactions on Ecology and the Environment, Vol 242, © 2020 WIT Press
www.witpress.com, ISSN 1743-3541 (on-line)
doi:10.2495/WP200071

States by 2030), the adaptation to climate change events (such as drought or flooding), and the likely increasing need to work with water from outside of EU borders [3]. The directives also aim to define what it means to have healthy water quality. Despite the established water infrastructures, sewage, industrial, and agricultural waste discharged into the waterways (diffuse pollution affects 90% of river basin districts, 50% of surface water bodies and 33% of groundwater bodies across the EU) continue to harm the environment or human health [4]. As a whole, they make the case that services need to provide for this good by protecting the transit of the water as well as the aquatic ecosystems and river basins the water comes from.

In this context, the risks to water security are framed strategically, e.g. what if the water gets cut off? In particular, vulnerabilities are framed not in humanitarian need or developmental progress, but as coming from external sources, such as extreme weather events or terrorist attacks, where standards and technology innovation are often looked to as mitigation measures [5], [6]. Similarly, solutions to these risks are framed in strategic technologies, like the use of remote sensors, smart technologies, and IoT systems. Within our work in the European Union Horizon 2020 funded project, aqua3S (grant agreement number 832876), we are currently addressing these challenges. The aqua3S project addresses seeks to create a water sensor system that utilises sensor technologies to support water safety. In doing so, IoT connected sensors, unmanned aerial vehicles (UAVs), satellite images and community generated social media observations on water quality will be used to identify anomalies in water networks. The collected information will then be presented to water network operators through an interactive user-face. Ultimately, these systems will only be used to assist decision making. In developing this system, robust physical and cyber-security measures are being developed, including encryption, privacy by design processes and access restrictions among others.

However, the relationship between water security, vulnerability, and human rights need to be fully appreciated in this context [2]. Movements like the Citizens' Initiative "Right2Water", reveal that even within these generally water secure regions, concerns still exist about how these definitions and solutions take into account differential access, equality, and uneven vulnerabilities. Exploring where interconnectivity and digitalisation in water infrastructures and strategic state solutions intersect with human rights and vulnerability offers an opportunity to explore the ethical dilemmas raised within water security. This article examines these water and cyber security frames to explore how to balance scales between innovation, inequal vulnerabilities and cyber-security threats. To approach the answer, it pays particular attention to demographics at risk of insecurity. Doing so, it sets the stage to create a reflexive framework for assessing the ethical implications of water security practices that help decision-makers see beyond normative and global assessments. We argue that to meaningfully understand the implications it is necessary to focus not only on where things can break but also on what is at stake, and for whom. The article bring into discourses about which bodies of water, pipelines, or ecologies could be affected (and by what) questions about how we know that the security provided is fair, just, and beneficial to all. It also highlights the important differences and overlaps between human security versus state security. The risk, at its core, is whether failing to develop a nuanced understanding of these relationships ultimately risks puncturing entry points for malign cyber infiltration into services that are essential for populations with the brunt of the potential resultant harms falling disproportionately on vulnerable and marginalised communities.

2 WATER SECURITY IN EUROPE

Within Europe, water security issues are defined through acceptable thresholds of threats and risks, including uncertainty, trade-offs, and social-economic and environmental externalities.

Running through these definitions around water security in legal frameworks across Europe are assumptions that the original water supply is relatively safe, protected against disease, is adequate and reliably availability for the needs, be they community, agriculture, or state [7]. This is reflected in European nations' massive investments of resources in decreasing water scarcity and similar high stressors. However, ethics and human rights – including how these definitions of security supports the necessities of a good life, human health, and ecosystem sustainability – are less well articulated in these definitions [8]. The underlying causes of water insecurity which have demonstrated discriminatory affects that are both less visible in the structure of these regulations and in water security practices themselves [9]. To help make these vulnerabilities more visible, questions like "is the water clean?" and "is the water reaching its destination?" need to be paired with "whose water is clean?" and "who is accountable for that protection?"

2.1 Unknowns and uncertainties as vulnerabilities

Such an approach requires understanding the vulnerabilities that need securing. Vulnerability is tied to a person's or community's ability to cope with a risk [10]. Coping abilities, however, are difficult to measure with data, numbers, or sensors. Moreover, quality of the data is only possible to ensure with knowns, yet vulnerabilities, often by definition, are the result of unknowns. This is exacerbated by intersecting practices of water security with state security where the daily acts of living (a child turning on the tap to wash hands in a kitchen sink pulling water from leaking lead pipes) become side-lined for equally important discussions around national water supplies, international relations necessary for such supplies, and trade-offs in different water usages (e.g. agricultural for food versus washing laundry at home) [11].

Indeed, measurements technologies are often weakest in areas with the highest uncertainties or variabilities [10]. Both new and legacy chemicals add to the chemical burden on Europe's populations and ecosystems, affecting public health in uncharted ways. Moreover, scientific uncertainty around what levels of some chemical substances are harmful has resulted in varying definitions of contamination. For example, the effects of the mobility, widespread use, and persistence of per-fluoroalkyl and polyfluoroalkyl substances (PFAS), which have been used for years in products from stain resistant textiles, Teflon, to pizza boxes are just now being made visible, resulting in contaminated drinking water and ubiquitous exposure throughout the Global North. Yet, despite government acknowledgement of long-term negative health effects of these and other micropollutants, there are no EU standards for drinking water on PFAS enabling use of such chemicals to go without restrictions, often without oversight [12], [13]. Further, for chemicals for which there exists guidance in the EU Drinking Water Directive, in some highly polluted areas concentrations of perfluorooctanoic acid (PFOA) and perfluorosulfonic acid (PFOS) in drinking water were well above the proposed limits [13]. Similar data gaps and asymmetries fall around diffuse water pollution from agriculture (DWPA), which can consist of pollutants from fertilizers and manure, sediment, and pesticides from farms, golf courses, private gardens and other rural domestic activities [14]. These pollutants have dispersed and less readily visible sources, making knowing where to put a sensor, when to sense, and what to sense for elusive. These uncertainties can be seen in the challenges in managing these chemicals in drinking water that emerge simply from their large numbers and variety, unconsolidated information, and limited studies on health impacts, particularly around persistence and regional sources that risk different types of exposure [15].

All too often guidelines are designed and applied only after contamination is discovered by other mechanisms, frequently through public outcry or new disease bubbles. While these mechanisms for monitoring keep increasing, validated methods are still lacking for some sources, like groundwater. Moreover, the policy and guidelines for making decisions around these issues lag, making it difficult for those facing insecurity from these arenas to know how to ask for change [12].

2.2 Differential and unequal vulnerabilities

Overall, low-income and minority communities disproportionately suffer from water pollution, are less able to afford treatment systems in home, lack technical and financial support, are more likely to live in areas with failing infrastructures and legacy chemicals like lead and have less resources for oversight [16]. However, water security discourse focuses on water infrastructure (e.g. water to buildings), and less about water within buildings or all populations. These differentials affect most those forgotten in society in general. For example, water security statistics miss the homeless who deal with a different type of complexity to water security, such as access to public toilets and water fountains [17]. More generally, such vulnerabilities are increasingly documented in many regions in a similarly wealthy nation, the U.S., as exemplified by the ongoing water crisis in Flint, Michigan and the water health crisis Hurricane Katrina and Hurricane Harvey.

Flint has demonstrated that many vulnerabilities are infrastructure based, both in home and in transit. The vulnerabilities faced were not results of the water source, per se, but result of the pipes themselves the water travelled through: in the lead leached from the pipes that travel to the kitchen faucets. It does not help that the particulate release from the pipes into the water was often sporadic and thus only sometimes registered toxicity when tested [16]. The families living in the houses fed by these pipes were most likely near or below the poverty level without the financial means to fix their pipes or the time to campaign for structural changes, reducing their visibly to decision-makers. These vulnerabilities are further masked by the young bodies that experienced the greatest harms, as they do not as readily have a political voice to push for change.

Climate stresses on water security are already a prominent discussion in policymaking. Nonetheless, even here, socio-economic disparities have the potential to influence the intensity of vulnerabilities. After Hurricanes Katrina and Harvey, the excess water from flooding meant facing toxicity residues in soils from activities that ended decades prior, leaking into groundwater and water sources undetected. In some cases, these previous activities were no longer recorded in land management or public housing planning. The lack of understanding around how these chemicals from the past affect water quality and health over the long term aggravates unknows and uncertainties for all. But, just as much, considering the trend to poorer, marginalised communities (often based in racial and class divides) living on former industrial and farm sites, such gaps in data and can have disproportionate effects. This cascading toxicity, moving from forgotten chemical legacies contaminating the ground to pollutants in drinking water had direct effect on the socio-economically disadvantaged populations that often live in or near former industrial sites [18]. By not collecting this data, structural water insecurity is built in for those already less powerful in society, trends that in the US have already instigated discourses around racial violence [11]. In the EU, these challenges are emerging in the discourses around toxic soils.

From the other direction, these kinds of power differentials also play out into who gets policed in relation to water security violations. Several studies have identified that a small subset of polluters cause the majority of the pollution, and that these polluters create disproportionate exposures to these chemicals through water for low income, non-white, and otherwise disadvantaged populations that don't have equal resources for making such activity visible to government decision-makers or for holding the companies accountable [19].

Much of the literature around the second phase of environmental justice has pushed this last point: justice requires mechanisms through which to make change, the procedures for participating in decisions and opportunities to build equality [20]. Security, from this perspective, is not just about the ability to obtain a good (e.g. clean drinking water) generally in society, but about how the risks to such a good are produced and distributed across society. This requires identifying positional inequalities that affect how vulnerabilities are spread in ways that disadvantage some while giving advantages to others, so that the basis for social and political change that support sustainable water security can emerge [20].

3 THE RELATIONSHIP BETWEEN CYBER SECURITY AND WATER SECURITY

The push towards the minimisation of operational costs and expenses has also led the charge towards technical innovation to better manage these increasingly evident forms of water insecurity. Amongst these is a drive to create increasingly extensive sensor networks across water systems to identify leaks and bursts, locate contaminants, and better manage disruptions, thereby creating new ways to see and eliminate uncertainties that lead to vulnerabilities [21]. In doing so, water service providers have looked towards the Internet of Things (IoT), utilising cloud computing to extend the reach and interconnectedness of their system [22]. This section maps these innovative potentials onto the vulnerabilities described in the previous section to explore how these efforts to protect water security produce new forms of benefits and vulnerabilities. It considers new risks that arise from cyber-attacks on these water systems in order to identify how cyber-security threats could manifest and influence how vulnerabilities and water security are framed.

Within the context of the water sector, this sensor innovation is most often achieved through connecting Supervisory Control and Data Acquisition (SCADA) systems – used for monitoring and controlling systems – with the IoT-cloud [22]. Beyond minimising costs, such remote sensor systems can capture issues that may otherwise be resource intensive and time consuming, such as the detection of pollutants and weak water pressure. Water service providers can produce a reactive and agile system that is able to detect anomalies and target threats in an efficient manner before actual harms manifest. Within analogue systems these anomalies may only be identifiable once a harm has already materialised or through the resource heavy system of manual testing, presenting concern to water services providers.

However, though these measures safeguard the integrity of the system, water security should not simply be addressed as a problem that can be solved with technical solutions. To begin with, cyber-attacks on smart water infrastructures in recent years demonstrate the vulnerability of critical infrastructure to infiltration. In addition, the ethical dimensions of such approaches to water insecurity need to be considered to inform our responses and provide an insight into how such responses shift our focus within the concept of water security in specific ways.

3.1 Cyber-attacks on water networks

Previous cyber-attacks on water networks offer some insights into key areas of vulnerability, the methods of attackers and the ways in which the effects can manifest. From the outset, it is important to note that the attacks outlined below do not necessarily relate to vulnerabilities arising out of IoT connected infrastructure. Nevertheless, they paint a useful picture of the vulnerability of systems that are able to be accessed remotely.

While these systems increase the nature and effectiveness of monitoring risks, they open these infrastructures to remote access and control. Moreover, they transform new forms of access to the pipelines and the water within into different vulnerabilities. To start, multiple attacks have occurred with worrying potential. Some are considered to have arisen out the offensive capacities of units specialising in cyber-attack that were not targeting water specifically but water as a means to state disruption. For instance, the Iranian Revolutionary Guard are said to have attacked the Bowman Avenue Dam in New York state, gaining control of the command and control system, with the ability to produce kinetic effects on the water system. While the flood gates were offline for maintenance and were therefore not accessible within the attack, this was fortuitous and demonstrates how vulnerable systems may actually be [23]. The 2016 attack on the "Kemuri" water plant (a pseudonym given to the plant), attributed to political hackers affiliated to Syria, gained control the levels of chemicals used within the water system, as well as retrieving large quantities of personal customer data [24]. In 2018, the Ukrainian security service reported that Kremlin-sponsored attackers targeted their water sanitation system with malware. This allowed the attackers to engage in espionage activities as well as kinetic damage to the water system. It was one of a suite of Ukranian systems affected – from transport, to governmental authorities, and even radiation monitoring systems at the Chernobyl Nuclear Power Plant [25]. Through these cyber-security attacks, water insecurity becomes a tool in geo-political conflict.

Outside of the geopolitical context, attacks are used to create disruptions for a range of reasons from financial gain to social activism. A remote attack at the Maroochy Water Services facility in Australia in early 2000, led to a loss of communication and pump control capabilities, altered the pump station configuration, and set off false alarms. In 2006, a cyber-attack on the Pennsylvania Water Filtering Plant in America gave attackers the ability to alter the concentration levels of disinfectants within the potable water. In another example from the Tehama-Colusa Canal in 2007, a former employee with had an intimate understanding of control systems accessed and damaged the computer system and diverted water to the local farms. In yet another, in 2019, governmental services, including water, in Florida were targeted in a ransomware attack. The attack compromised Riviera Beach Water Utility's computer systems, preventing them from using their pumping stations, water quality testing functions and their payment operations [23].

The diversity of motive, nature and scope of attacks demonstrates a need to be attuned to the possibility of attack across the entire cycle of the water system, materialising in an array of harms. Water insecurities emerge not from the lack of water delivery systems or ability to clean water, but from cross-border political dynamics to malicious actor seeking financial gains. A full understanding of system and human vulnerabilities must therefore be understood and built into the decision-making process, planning and response measures.

3.2 Key cyber vulnerabilities in relation to water

So how do the impacts that have manifested in previous attacks translate into real human harm? Cyber-attacks can result in kinetic effects on critical features of water services, such as the manipulation of flood gates, interference with chemical and water levels and even the diversion of irrigated water. Moreover, where customer information is contained in water systems – the type necessary to get the water from plant to house – it can be remotely retrieved, posing a very real concern for the privacy rights of service users. A distinct set of threats may appear in contrast to the usual anticipated harms of lack of water provision.

As technological innovation is frequently built into existing legacy systems, it is necessary to ensure that retrofitting innovation does not ignore the concern that legacy systems may include outdated systems vulnerable to infiltration. Where unpatched code remains, innovation measures must seek to actively address and overcome these challenges [26]. SCADA systems which integrate old and new technologies, – industrial business systems and the IoT-cloud system – they become more susceptible to infiltration than the traditional, less advanced SCADA systems [22]. Whilst recognising the advantages that IoT connected systems may bring, they may also include a number of vulnerabilities. These include configuration errors from default factory settings, vulnerability in cloud services, memory corruption and weakness in validating input data, and ultimately the vulnerability of system commands and information to interference [22].

The cumulative result of these vulnerabilities are that the combined integrated systems are at risk of advanced persistent threats; the lack of data integrity where data is destroyed; man-in the middle attacks where the attackers gained illegitimate access or monitors the messages and activities within the system; replay attacks which delay messages sent to physical devices and denial of service attacks which prevents the system from performing tasks by overloading the computer resources [22].

In order to chart the levels and nature of cyber-vulnerability, a focus on the underlying structural issues that may lead to such vulnerability is key. Here, a drive to reduce costs appears to be a motivator to enhance remote sensing and control capabilities of water infrastructures. Does this drive results from a profit focus to drive down costs at the expense of the community and result from an underfunding of the water sector? Understanding why such tools are in place can point to what kinds of mitigation measures are needed to reduce these new risks. Some could continue to be technical, like regular system or sensor updates. Some are political or organisational, including staff training to avoid human error, new regulations to manage silent polluters, or increased government funding [27]. Some require resources for the immediate moment, and others require resources throughout a longer water or pollutant lifecycle. Sufficient attention needs to be paid to the continuing costs necessary to upkeep the cyber security framework. The development of such systems is not a one-off event, and the failure to update both the technology and the socio-economic and political systems that support them may mean that they do not keep pace with the advancement and innovation of malicious threats, reopening water networks to potential harm [26].

Moreover, despite the increasing frequency of cyber-attacks on water networks, the concept remains relatively novel, particularly within the European context. As such, where remote sensing is utilised within the system, the potential for cyber-attacks may not receive adequate attention or investment. In this sense, where the resources are scarce in the first instance, water service providers may be hesitant to invest the necessary funds to safeguard against a vague hypothetical and potentially unrealised threat. As a result, the security concerns must be appreciated by senior staff, or those with decision and investment-making positions, in order to ensure that they are aware of the seriousness of these threats [27].

Technical and organisational measures can be developed to better protect connected water systems. But just as importantly, each new measure put in place to reduce water insecurity shifts the focus and aim of water security. Paul Rosenberg, the mayor for the district in which the compromised Brookman Avenue Dam was located, responded to the attack by taking the dam controls offline, holding that "the risks outweigh the benefits" [28]. As sensors bring geopolitics into view, the needs for more resources shift towards cyber security (away, likely from more marginalised social needs), and focuses water security as an infrastructure problem. Yet, invisible toxicities bring into focus how living conditions and structural inequalities in society, even in wealthy nations, still drive less tangible and democratic water insecurities, vulnerabilities only partially addressed through sensors. How, then, can we use the lens of vulnerability to better understand what benefits and harms are created by the different measures developed to improve water security?

4 IMPACT ASSESSMENTS

This section outlines ways to sensitise water service providers, policy makers, and technology designers to the ethical and human rights challenges of water security, and how to better assess the proportionality of the risks and benefits in their security frames. Drawing on the methodology of an Ethical and Privacy Impact Assessment, it provides a structured and reflexive approach for identifying and assessing risks, as well as developing recommendations to be considered and actioned where possible within system innovation and development. Within the European Union, Privacy Impact Assessments and Data Protection Impact Assessments are often compulsory under the General Data Protection Regulation in order to demonstrate compliance with the legal and regulatory requirements. Such practices are now widely used across the globe [29]. Increasingly, similar processes are being used to capture a broader array of considerations, including ethical impact assessments and societal impact assessments [30], [31].

Building from an E/PIA provides a distinct avenue to include conversations around individual and diverse community insecurity into technology and policy decisions, thereby attuning decision-makers to these considerations so they can recalibrate their ideas to their potential real-world impacts on diverse populations and any disproportionate impacts on individual's rights [32]. To support such a process, we propose a set of questions, based on the themes, overlaps, and disparities between the discourses in western water insecurities and water security solutions we present here.

These questions have no right answers but help make visible and transparently engage with what may otherwise be morally opaque at first glance [33]. They require a reflexive ability to grapple with the particulars of each setting, porous to the dynamic range of issues. By starting with such questions for an E/PIAs framework, they can contribute towards informed decision-making, protection of societal concerns, and overall effective risk management strategy [34]. These questions intend to uncover some of the key vulnerabilities of populations and indeed the water sector as a whole.

1. What are the fundamental vulnerabilities to water insecurity that the innovative system is trying to protect and how does the innovation prevent or reduce these vulnerabilities?

2. To what extent is information on the local population disaggregated to include characteristics such as gender, age, class, disability to appreciate their particular vulnerability and resilience to water insecurity?

3. How do these characteristics influence how individuals/communities interact with the water system?

4. How do water system's response measures seek to protect a particular area, sector, site, or community at the expense of another? If so, how is this prioritisation calculated and what are the potential human impacts of this prioritisation?

5. What are the human impacts if a community or responsible authority have sufficient resources to take advantage of the solution?

6. To what extent do the processes used to identify and evaluate risks to the water network also contain information on the vulnerability and resilience of the population against such risks?

7. What are the human impacts that can arise from cyber-attacks? Does the potential for these impacts out-weight the existing water security threats that the system is trying to mitigate (taking into account the specific harms/burdens per community outlined in Question 2)?

8. To what extent are there sufficient resources for training personnel and updating systems to safeguard against the new vulnerabilities arising through the solution?

9. Have compounding harms that arise for the simultaneously disruption of other critical infrastructures and socio-economic activities been mapped?

10. How does the system adjust to accommodate new knowledge regarding potential threats to communities or individuals (e.g. increased awareness on new forms of pollutants and new forms of cyber-attack)?

11. What are the mechanisms for policy change or accountability to address and change the root causes of water insecurity?

5 CONCLUSION

This paper adopts an anthropocentric view of the impacts of water insecurity. Nevertheless, the proffered impact assessment model can be tailored to other referent objects, such as the environment or agriculture, recognising both these objects' innate worth and need for protection from pollution and the complexity of how to define and articulate risks of water-based harms. Ultimately, such an approach provides for a holistic understanding of water security making it possible to anticipate and prevent harms from occurring in the first place. With such a perspective, mitigation measures for water pollution can be better designed to addresses underlying causes and drivers, and in doing so informs effective and responsive measures and responses.

Putting vulnerabilities into conversation with technological solutions makes visible how water security is a right that looks beyond the flow of water between source to a view that interconnects international, local, human, environmental, economic, and political concerns [4]. Understanding the social, economic, political context of technology adoption, particularly the complexities barriers, can make visible the indirect impacts that have differential effects yet shape both the concept of water security as well as the vulnerabilities that water security addresses [35].

The vulnerabilities that arise with increasing innovation in the water sector have resulted in a recalibration of how we look at the threat of water insecurity in the European context. They show how a new technological solution to one problem (networked sensors to detect chemical threats, a common and diversely experienced water vulnerability at the community and individual scape) can shift the view to a state and political sense of security (facing new threats from geo-political dynamics). As the threat of cyber-attack looms larger, we must be aware of how humans can be affected in their unique contexts, the new drivers of harm (for instance, as geo-political dynamics are introduced into the critical infrastructure ecosystem), and how we should prioritise our resources to address these threats. At its core, increasing innovation in the water sector demands a clear understanding of the proportional risks, based in an appreciation of marginalised people and communities and capable of ensuring that their voices and concerns are factored in responses.

These particular concerns can be better identified and addressing by engaging in a meaningful ethical impact assessment process. Starting from reflexive questions, ones without rights or wrongs but that pose dilemmas or interrogate assumptions, can make visible to all what goes into a water security measure; and as importantly, what and who is left out.

They make it possible to better articulate why decisions get made, understand what kinds of policy pathways are necessary for supporting change, help identify new risks that might emerge, and make visible what might be otherwise further masked.

Fundamentally, when considering how water pollutants should be measured and addressed, we must steep this understanding of the human context. A pollutant detection system will ultimately be human-agnostic, measuring chemical makeup without reference to the fact that the resultant water insecurity risks are not equally distributed across societies. Indeed, new technologies to monitor pollution can even reshape what risks water pollution potentially poses. The adoption of an impact assessment can help to bring in this contextual analysis. Moreover, as highlighted within Section 2, pollutant-based harms occur on an intersection of cross-cutting behaviours, events and structural dynamics that dictate harms' scope and severity. The impact assessment methodology can help to develop an appreciation of how innovations interact with these relationships and other harms that arise from separate areas (such as privacy). Finally, developing this intersectional insight can bring community voices into the equation. In this respect, it is possible to complement current expert-led approaches to water insecurity with bottom-up participatory measures. Consequently, our understanding of water pollution can develop a well-rounded picture of its context, which in turn will help both assess and prioritise innovations and policy, as well as tiers of harms.

REFERENCES

[1]   United Nations, UN Analytical Brief: Water Security and the Global Agenda, 2013. https://www.unwater.org/app/uploads/2017/05/analytical_brief_oct2013_web.pdf.
[2]   Maganda, C., Water security debates in 'safe' water security frameworks: Moving beyond the limits of scarcity. *Globalizations*, **13**(6), pp. 683–701, 2016.
[3]   Security Research Community of Users, Water Security and Safety, CoU Brief, no. 2, Mar. 2018.
[4]   Scocca, G., Strengthening international water security: The European Union's proposal. *World Water Policy*, **5**, pp. 192–206, 2019.
[5]   Zeitoun, M., *The Web of Sustainable Water Security, in Water Security: Principles, Perspectives and Practices*, Routledge: London, pp. 11–25, 2013.
[6]   EurEau, The Need for Greater EU Policy Coordination Realising the Water Framework Directive, 16 May 2017. http://www.eureau.org/resources/position-papers/140-greater-eu-policy-coordination-may2017/file.
[7]   Wouters, P., Water Security: Global, regional and local challenges, Institute for Public Policy Research (IPPR), 2010.
[8]   Bakker, K. & Morinville, C., The governance dimensions of water security: a review. *Philosophical Transactions of the Royal Society A: Mathematical, Physical and Engineering Sciences*, **371**, 2013. http://doi.org/10.1098/rsta.2013.0116.
[9]   Vörösmarty, C., et al., Global threats to human water security and river biodiversity. *Nature*, **467**, pp. 555–561, 2010.
[10] Garrick, D., et al., Water Security, Risk and Society – Strategic Report on Research Findings, Gaps and Opportunities, Submitted to the Economic and Social Research Council by Oxford University Water Security Network, 2012. www.water.ox.ac.uk.
[11] Dillon, L. & Sze, J., Police Power and particulate matters: Environmental justice and the spatialities of in/securities in U.S. Cities. *English Language Notes*, **54**(2), 2016.
[12] Cordner, A., et al., Guideline levels for PFOA and PFOS in drinking water: the role of scientific uncertainty, risk assessment decisions, and social factors. *Journal of Exposure Science and Environmental Epidemiology*, **9**, pp. 157–171, 2019.

[13] European Environment Agency, Emerging chemical risks in Europe -PFAS, Briefing no. 12/2019, 2019. https://www.eea.europa.eu/themes/human/chemicals/emerging-chemical-risks-in-europe.

[14] Graversgaard, M., et al., Opportunities and barriers for water co-governance—a critical analysis of seven cases of diffuse water pollution from agriculture in Europe, Australia and North America. *Sustainability MDPI*, **10**(5), pp. 1–39, 2018.

[15] Guelfo, J.L. et al., Evaluation and management strategies for per- and polyfluoroalkyl substances (PFASs) in drinking water aquifers: perspectives from impacted U.S. Northeast communities. *Environmental Health Perspectives*, **126**(6), 2018.

[16] Katner, A., et al., Weaknesses in federal drinking water regulations and public health policies that impede lead poisoning prevention and environmental justice. *Environmental Justice*, **9**(4), pp. 109–117, 2016.

[17] Hale, M., Fountains for environmental justice: Public water, homelessness, and migration in the face of global environmental change. *Environmental Justice*, **12**(2), pp. 33–40, 2019.

[18] Boudia, S. et. al., Residues: Rethinking chemical environments. *Engaging Science, Technology and Society*, **4**, pp. 165–178, 2018.

[19] Collins, M., Munoz, I. & Jaja, J., Linking "toxic outliers" to environmental justice communities. *Environmental Research Letters*, **11**(1), 2016.

[20] Curran, D., Environmental justice meets risk-class: The relational distribution of environmental bads. *Antipode*, **50**, pp. 298–318, 2018.

[21] Sammaneh, H. &. Al-Jabi, M., IoT-enabled adaptive smart water distribution management system. *International Conference on ICPET Promising Electronic Technologies (ICPET)*, pp. 40–44, 2019.

[22] Sajid, A., Abbas, H. & Saleem, K., Cloud-assisted IoT-based SCADA systems security: A review of the state of the art and future challenges. *IEEE Access*, **4**, pp. 1375–1384, 2016.

[23] Hassanzadeh, A., et al., A review of cybersecurity incidents in the water sector. *Journal of Environmental Engineering*, **146**(5), 2020.

[24] Tsagourias, N., Cyber attacks, self-defence and the problem of attribution. *Journal of Conflict and Security Law*, **17**(2), pp. 229–244, 2012.

[25] Martin, A., Russian hackers targeted Ukraine's water supply, security service claims, *Sky News*, 11 Jul. 2018. https://news.sky.com/story/russian-hackers-targeted-ukraines-water-supply-security-service-claims-11432826. Accessed on: 17 Apr. 2020.

[26] Adepu, S., et al., Investigation of Cyber Attacks on a Water Distribution System, ArXiv abs/1906.02279, 2019.

[27] Germano, J. H., Cybersecurity Risk and Responsibility in the Water Sector, American Water Works Association, 2018.

[28] Esposito, F., Westchester village finds clever solution to thwart hacking of critical infrastructure, *Rockland/Westchester Journal News*, 8 Jan. 2020. https://eu.lohud.com/story/news/local/westchester/rye-brook/2020/01/08/iran-hacked-rye-brook-dam-2013/2846127001/. Accessed on: 17 Apr. 2020.

[29] Wright, D. & Friedewald, M., Integrating privacy and ethical impact assessments. *Science and Public Policy*, **40**(6), pp. 755–766, 2013.

[30] Wright, D., Ethical impact assessment. *Ethics, Science, Technology and Engineering: A Global Resource*, eds J. Holbrook & C. Mitcham, 2nd ed., Macmillan Reference: Farmington Hills, 2015.

[31]   Kush Wadhwa, K., Barnard-Wills, D. & Wright, D., The state of the art in societal impact assessment for security research. *Science and Public Policy*, **42**(3), pp. 339–354, 2015.
[32]   Carroll, J.M., Five reasons for scenario-based design. *Interacting with Computers*, **13**, pp. 43–60, 2000.
[33]   Brey, P., Disclosive computer ethics. *Computers and Society*, **30**(4), pp. 10–16, 2000.
[34]   Kloza, D et al., Data Protection Impact Assessments in the European Union: Complementing the new legal framework towards a more robust protection of individuals, d.pia.lab Policy Brief No 1/2017, 2017.
[35]   McDonnell, R. et al., Water Security, Risk and Society – Knowledge Exchange Opportunities for UK and European Agencies, Briefing note submitted to the Water Security Knowledge Exchange Programme by Oxford University Water Security Network, 2012. www.water.ox.ac.uk and www.wskep.net.

SECTION 3
COASTAL AND OFFSHORE
POLLUTION

ECOLOGICAL AND MOLECULAR APPROACH TO THE ASSESSMENT OF OIL POLLUTION: A COMPARATIVE STUDY BETWEEN TWO COASTAL MARINE (MEDITERRANEAN AND PATAGONIAN) ECOREGIONS

GIUSEPPE ZAPPALÀ[1], GABRIELLA CARUSO[2], RENATA DENARO[3],
FRANCESCA CRISAFI[1] & LUIS SALVADOR MONTICELLI[4]
[1]National Research Council, Institute of Biological Resources and Marine Biotechnologies (CNR-IRBIM), Italy
[2]National Research Council, Institute of Polar Sciences (CNR-ISP), Italy
[3]National Research Council, Water Research Institute (CNR-IRSA), Italy
[4]National Research Council, Institute for Coastal Marine Environment (CNR-IAMC), Italy

ABSTRACT

Microorganisms are considered as sentinels of environmental changes, but microbial response to the presence of contaminants, such as hydrocarbons, is still not fully understood. This study aimed at assessing the response of the microbial community to the presence of oil pollution, by comparison of two ecosystems, Milazzo Gulf (Sicily, Italy) and Caleta Cordova (Argentina), as representative of two different temperate Mediterranean and cold-temperate Patagonian ecoregions, respectively. Water and sediments were sampled at coastal stations characterized by different levels of hydrocarbon contamination and analyzed for their microbial enzymatic activity rates; the presence of hydrocarbon-degrading bacteria was also determined. The study approach allowed to determine that microbial metabolism was significantly reduced at the polluted stations, suggesting the detrimental effects of contaminants on organic matter degradation process. The isolation of strains of hydrocarbon-degrading bacteria confirmed that the oil contamination favoured the growth of strains specifically adapted to metabolize hydrocarbons and actively involved in the remediation processes.
Keywords: oil pollution, microbial response, enzyme activities, hydrocarbon-degrading bacteria.

1 INTRODUCTION

Oil pollution is a major threat for world's oceans and a severe cause of environmental deterioration [1], [2]; petroleum hydrocarbons can enter into aquatic environments as a consequence of natural oil sources following freshwater runoff, rivers and sewage systems [3], [4] or as a result of human activities related to oil extraction and refinery, causing chronic pollution (i.e. maritime transports or shipping activities) or following occasional oil spill events caused by pipeline failures or shipping disasters (i.e. Exxon Valdez, Deepwater Horizon) that generally result in acute pollution.

In order to optimize decision making response to oil spills, a deep knowledge of the fate of oil spill and ecological effects related to hydrocarbon compounds discharged into the environment is necessary. In temperate regions, an early weathering phase of oil spills leads to its evaporation, absorption in water and dispersion [5]. Following hydrodynamic forcings such as currents and waves, emulsification and dispersion of spilled oil starts, making its components available as a carbon source to a fraction of indigenous microbiological populations able to metabolize them [1], [6], [7]. Hydrocarbon-degrading bacteria are autochthonous microorganisms widespread in the marine environment, able to use hydrocarbons as the sole carbon source and therefore extremely specialized in the removal of these contaminants [1], [8]. Due to their persistence and accumulation along the trophic web, the toxic effects of hydrocarbon contamination on marine biota, such as benthic communities are well known [2], [9]. Knowledge of the effects of oil pollution on the microbial community structure and function is comparatively low.

WIT Transactions on Ecology and the Environment, Vol 242, © 2020 WIT Press
www.witpress.com, ISSN 1743-3541 (on-line)
doi:10.2495/WP200081

Microorganisms are considered as sentinels of environmental changes; indeed, they are able to respond to the presence of contaminants by modifying their abundance and/or metabolism [10]. Heterotrophic bacteria, which play a major role in the particulate to dissolved matter transformation, are known to promptly respond to environmental changes, modifying their metabolic patterns according to available organic compounds. Thanks to their metabolic activities on organic polymers (proteins, organic phosphates), marine microorganisms make the nutrients (N, P) necessary for the development of hydrocarbon bacteria available in inorganic form. The addition of nutrients is in fact a strategy used to stimulate the bioremediation process, as the *in situ* microbial degradation of contaminants is generally slow due to the low availability of nutrients and oxygen. Therefore, the estimate of the decomposition rates of organic matter can provide useful information on the flux of elements made available by the microbial activity of enzymatic hydrolysis; this can be a biostimulating factor towards hydrocarbon bacteria. *In situ* characterization of metabolic potential is therefore essential to assess the potential ability of microbial assemblage to be involved in decomposition processes. Indeed, microbial enzymatic activities can constitute potential bio-indicators of the microbial degradation capacity on hydrocarbons and an alternative approach to monitor the progress of the processes of bioremediation [3].

The rationale of this study was to investigate the *in situ* potential of autochthonous hydrocarbon-degrading bacteria during the natural occurring clean-up processes in two cold and temperate environments which are at risk of oil pollution, Caleta Cordova because of the oil-based industrial activity, Milazzo Gulf due to the presence of a refinery plant and intensive maritime traffic within its harbor. To this aim, within the joint Cooperation Agreement between Italy and Argentina funded by CNR, a study was undertaken on the pelagic and benthic compartments in order to assess the response of the microbial community to the presence of oil pollution, by comparison of two ecosystems, Milazzo Gulf and Caleta Cordova, as representative of temperate Italian and cold-temperate Patagonian ecoregions, respectively.

The specific objectives of this study were: (i) to assess the microbial extracellular enzymatic activity of the microbial assemblage in the water and sediments of coastal marine sites differently impacted by oil pollution; and (ii) to isolate hydrocarbon-degrading bacteria from the sediments. The study approach allowed to determine the expression of microbial activities related to organic matter turnover and metabolic functioning simultaneously to the characterization of the microbial assemblage specifically involved in bioremediation process.

2 MATERIALS AND METHODS

2.1 Study sites and sampling

Samplings of water and sediments were performed at Mediterranean (Milazzo Gulf) and Patagonian (Caleta Cordova) coastal marine stations. The location of the study areas with indication of the sampling stations is shown in Fig. 1.

2.1.1 Italian site
The Gulf of Milazzo ecosystem (Fig. 1(a)) is located in the Tyrrhenian Sea, along the north-eastern coast of Sicily; in its western part it receives organic matter inputs coming from stream and urban and industrial settlements. A study by Yakimov et al. [11] reported concentrations of alkanes (C12–C16 atoms) and alkyl phthalates of 143 and 61 μg/kg per dry weight of sediment, respectively. Unlike the hydrological and general biological parameters,

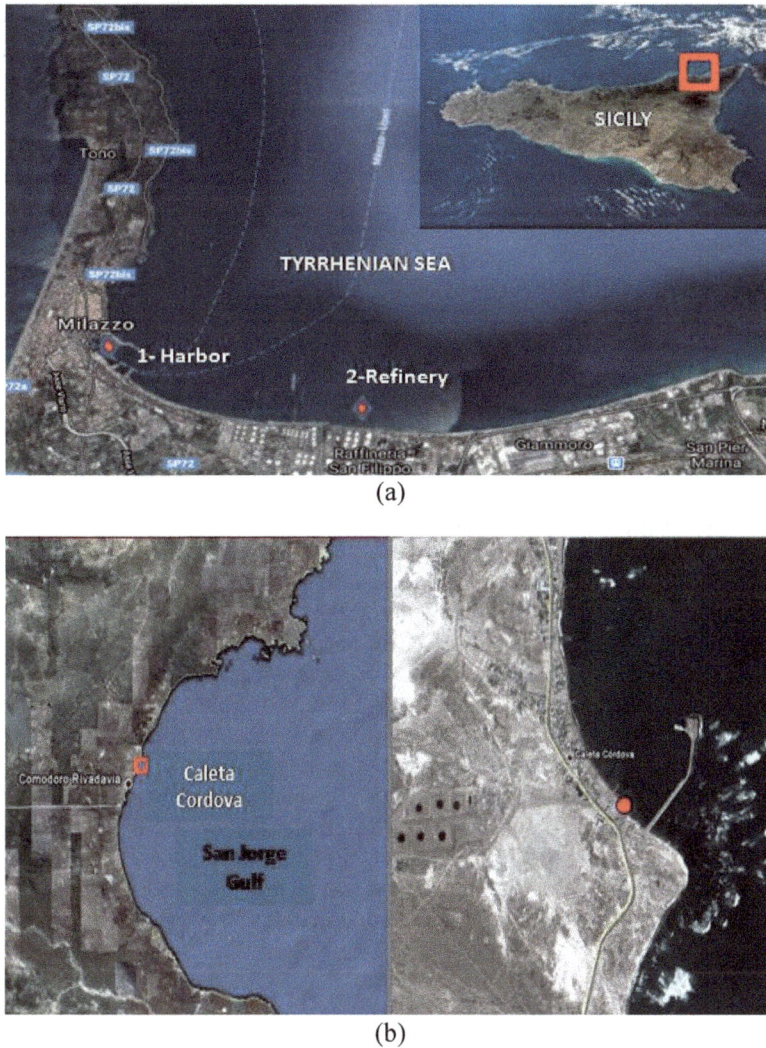

(a)

(b)

Figure 1:   Location of the study areas. (a) Milazzo Gulf as a temperate Mediterranean site; and (b) Caleta Cordova as a sub-Antarctic Patagonian site. The sampling sites are indicated by the red symbols.

bacterial dynamics and metabolism in this ecosystem remain poorly understood. On 3 December 2009, two stations – one located within the Milazzo harbor and one in front of the Milazzo refinery – were sampled.

2.1.2 Patagonian site

Sampling in Patagonia was performed on 8 June 2010, in a seasonal period (late autumn–winter) comparable with that in which the activities were carried out in Italy. The sampled site was Caleta Córdova (Province: Chubut) located on the San Jorge Gulf, within the Comodoro Rivadavia oil basin. The chosen area represents one of the centres of the Argentine oil industry characterized by the presence of numerous augers near the coast. In coastal sediments, total aliphatic and aromatic hydrocarbons reached concentrations of 106.3 and 107.8 µg/g per dry weight of sediment, respectively [12]. Two stations 100 m far from each other – indicated as CC10(1) and CC10(2) – were sampled. Sediment sampling was carried out using plexiglass pipes with sharp edges, collecting portions of sediment while remaining within a circle of about 1 m^2.

2.2 Measured variables

At the time of sampling, *in situ* measurements of temperature, pH and oxidation-reduction potential Eh were carried out. The following microbial parameters were measured: extracellular enzymatic activities (EEA: leucine aminopeptidase, LAP and alkaline phosphatase, AP) and heterotrophic bacterial abundance (MA) by Marine Agar plate counts. EEA were determined using the specific fluorogenic substrates L-leucine-7-amido-4-methyl-coumarin hydrochloride (LEU-MCA), 4-methylumbelliferyl (MUF)-phosphate for LAP and AP) and reported in terms of the maximum velocity of hydrolysis (V_{max}) [3].

2.3 Hydrocarbon-degrading bacteria isolation

On the same day of sampling, the collected seawater and sediments samples were mixed to make them homogeneous and distributed among the researchers according to the needs of the planned analyses. Microcosms were set up in order to enrich the relative densities of hydrocarbon-degrading bacteria formed by the natural sample slurry, amended with nutrients-containing ONR7a medium and hydrocarbons as sole source of carbon. Samples collected at Caleta Córdova (austral winter season) were incubated at 4°C according to the *in situ* measured temperature (Table 1) for approximately 5 weeks. For the samples collected in the Gulf of Milazzo incubation lasted 15 days. The microcosms prepared in 1/1 ratio with ONR7a were added with the substrate (0.1% tetradecane or 0.1% Phenanthrene). The microcosms were prepared in triplicate for each site: the first set was designed for bacterial isolation and kept at 4°C, the second one was kept at 15°C, the third one – that was immediately treated with hydrochloric acid and stored at 4°C – was used for the chemical analyses.

2.4 Phylogenetic analyses

Analyses of the 16S rDNA gene sequences of isolates were performed as previously described [13]. Briefly, DNA extraction of bacterial isolates was performed with the CTAB method [14]. 16S rDNA loci were amplified using 16S rDNA forward domain-specific bacteria, Bac27_F (5'-AGAGTTTGATCCTGGCTCAG-3') and reverse primer Uni_1492R (5'-TACGYTACCTTGTTACGACTT-3') [15]. The amplification reaction was performed in a total volume of 50 µl mixture containing solution Q (Qiagen, Hilden, Germany), Qiagen reaction buffer, 1 µM of each forward and reverse primer, 10 µM dNTPs (Gibco, Invitrogen Co., Carlsbad, CA, USA), 2.0 µM (50–100 ng) of template and 2.0 U of Qiagen Taq Polymerase (Qiagen). The amplified 16S rDNA was sequenced using Macrogen Service (Korea). SIMILARITY_RANK from the Ribosomal Database Project (RDP) [16] and FASTA Nucleotide Database Query were used to determine partial 16S rDNA sequences to estimate the degree of similarity to other 16S rDNA gene sequences. Analysis and phylogenetic affiliates of sequences was performed as previously described [17].

2.5 Chemical analyses

The selection of hydrocarbon-degrading bacteria was carried by estimation of degradation efficiency within microcosms according to Crisafi et al. [13] by the analysis of total extracted and resolved hydrocarbons (TERHC). Briefly: at the fixed time points, TERCH were extracted from water sample following the 3510C EPA procedure (US Environmental Protection Agency). 500 ml of seawater were added with a mixture of acetone-dichloromethane 1:1 (vol/vol) and shaken at 150 rpm for 30 min; the extract was treated with anhydrous Na_2SO_4 (Sigma-Aldrich, Milan, Italy). The same treatment was repeated twice with 500 ml of dichloromethane. After drying, residues were re-suspended in 1 ml of dichloromethane and analyzed by gas chromatography [18] according to Crisafi et al. [13]. Data obtained were normalized by decamethylanthracene (internal spike/control of biodegradation) and used to evaluate the relative biodegradation of hydrocarbon.

3 RESULTS

3.1 Physical–chemical variables

3.1.1 Milazzo Gulf
The values of the physical and chemical variables measured in Milazzo Gulf are shown in Table 1. Temperature showed higher values at the refinery station compared to the harbor station. pH and Eh were in a range similar to other temperate marine environments.

Table 1: Milazzo Gulf: Physical–chemical variables measured at the sampling sites.

Site	Milazzo Gulf		
	Temperature	pH	Eh
1. Harbor	17.21	7.8	144
2. Refinery	18.32	7.6	168

3.1.2 Caleta Cordova
Physical–chemical measurements performed at Caleta Cordova gave the dataset reported in Table 2. Temperature values were typical of a cold sub-Antarctic region and did not differ significantly between the two stations; pH and Eh ranges also did not show evident differences. Despite the proximity to the extraction plants, no evident signs of a contamination from oil or hydrocarbons were observed, either due to halos on the water surface or to residues on the beach. Some patches of the collected sediment samples showed parts of dark sand, probably due to anaerobic zones rather than to the presence of hydrocarbons.

3.2 Enzyme activity rates

In both the examined sites, LAP predominated compared to AP, suggesting a higher amount of degradable organic polymers with a proteinaceous nature with respect to organic phosphates.

In the waters of Milazzo (Fig. 2), LAP activity rates ranged from 4.68 to 6.63 nmol/l/h, while AP showed activity rates comprised between 3.85 and 18.17 nmol/l/h.

In the sediments of Milazzo (Fig. 3), higher enzymatic levels were measured, ranging from 9.10 to 32.33 nmol/g/h and from 6.81 to 21.18 nmol/g/h for LAP and AP respectively.

Table 2:   Caleta Cordova: Physical–chemical variables measured at three points per each site.

Sampling site	Triplicated points	Temperature	pH	Eh
Caleta Cordova 1	1a	2.9	7.4	158
45°45.135'S	1b	3.1	8.0	165
67°22.341'W	1c	3.3	7.9	163
Mean ± s.d.		3.1 ± 0.2	7.8 ± 0.3	162 ± 4
Caleta Cordova 2	2a	3.1	7.9	140
45°44.911'S	2b	3.6	8.0	146
67°22.597'W	2c	3.7	7.9	170
Mean ± s.d.		3.3 ± 0.3	7.9 ± 0.1	152 ± 16

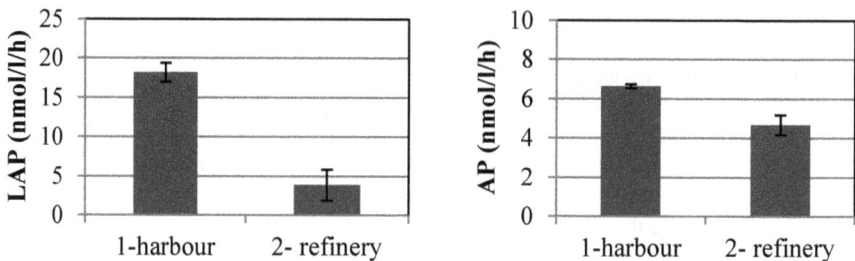

Figure 2: Mean ± s.d. enzyme activity rates measured in the Milazzo Gulf water.

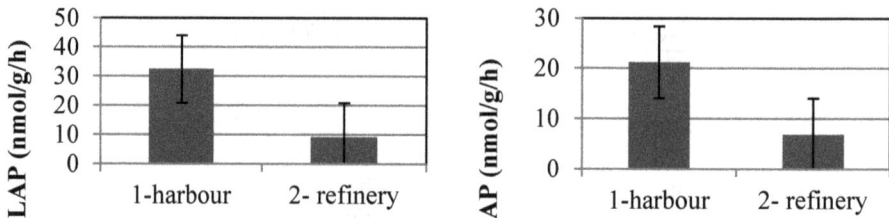

Figure 3: Mean ± s.d. enzyme activity rates measured in the Milazzo Gulf sediments.

Higher metabolic activity levels were always recorded inside the Milazzo harbor, while in front of the refinery the metabolism of the microbial community inhabiting both pelagic and benthic compartments appeared to be significantly ($P < 0.01$) reduced.

In Caleta Cordova LAP values ranged from 2 to 15 nmol/l/h and from 901.81 to 1568.66 nmol/g/h in water and sediments (Figs 4 and 5), respectively. In the same compartments, AP values were comprised between 0.21 to 1.49 nmol/l/h and between 56 and 141.3 nmol/g/h, respectively. The spatial distribution patterns of both LAP and AP enzymatic values showed a greater metabolic activity of the microbial community at station 2 both at the water and sediment level.

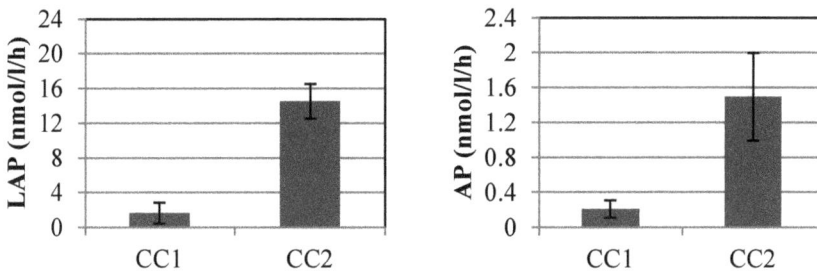

Figure 4: Mean ± s.d. enzyme activity rates measured in the Caleta Cordova water.

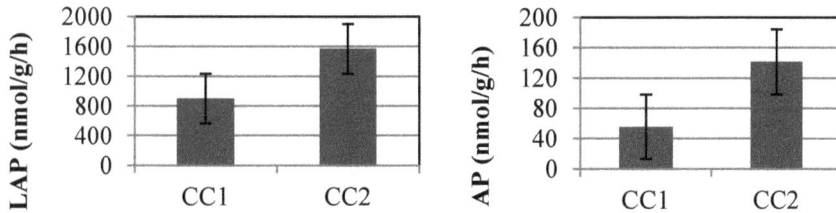

Figure 5: Mean ± s.d. enzyme activity rates measured in the Caleta Cordova sediments.

3.3 Hydrocarbon-degrading bacteria

After 8 days of incubation at the established temperatures, 100 µl aliquots of the slurry collected from microcosms were used for serial dilutions up to a dilution factor of 10^{-4}. 100 µl of each dilution were inoculated in ONR7a amended with previously used hydrocarbons, namely 0.1% tetradecane or 0.1% phenanthrene (w/v) collected from Caleta Cordova and maintained at 4°C, while samples collected from Milazzo Gulf were maintained at 15°C. When an increase in turbidity was observed in the cultures, aliquots of 100 µl were inoculated in ONR7a solid medium. As soon as the different morphologies of colonies were distinguishable, 10 strains from Caleta Cordova samples and 23 strains from Milazzo Gulf were selected for further isolation step and then phylogenetic analyses.

Results of the isolation of bacterial strains are reported in Tables 3 and 4 for Milazzo Gulf and Caleta Cordova, respectively. Bacterial strains isolated from Milazzo Gulf showed in general a higher degradation performance, also due to the higher experimental temperature. All the strains grown in tetradecane already after 20 days showed a decrease of the pollutant in the range of 50–60%. A peculiar behavior was observed in *Alcanivorax dieselolei B5*, a hydrocarbonoclastic bacterium well-known for the capability to degrade oil, also by the production of biosurfactants [19]. The strain isolated from Milazzo Gulf showed the highest performance, degrading in 20 days the 80% of tetradecane. Strains belonging to *Marinobacter* and *Halomonas* genera degraded the 60% of the pollutant. The cultures supplemented with phenanthrene were maintained for 30 days, obtaining a 25% decrease of the total amount of the pollutant in presence of *Pseudomonas*, and lower percentages (15–20%) in presence of *Rhodococcus* and *Vibrio*.

Table 3: Milazzo Gulf: Number of bacterial isolates obtained on media containing tetradecane and phenantrene as substrates.

Number of isolates	Closest match to	%	Accession number	Substrate
7	*Alcanivorax dieselolei B5*	98	NR_074734	Tetradecane
5	*Marinobacter sp. PJ-24*	99	KC200265	Tetradecane
3	*Halomonas venusta*	98	KJ416384	Tetradecane
3	*Rhodococcus yunnanensis*	97	JX827199	Phenantrene
4	*Pseudomonas sp. PAHAs-1*	97	KF483151	Phenantrene

Table 4: Caleta Cordova: number of bacterial isolates obtained on media containing tetradecane and phenantrene as substrates.

Number of isolates	Closest match to	%	Accession number	Substrate
4	*Marinobacter antarcticus*	98	NR_108299	Tetradecane
3	*Oleispira antarctica*	99	NR_025522	Tetradecane
1	*Rhodococcus sp. ice-oil-488 s*	98	DQ521396	Tetradecane
2	*Sphingopyxis flavimaris*	99	NR_025814	Phenantrene
4	*Pseudoalteromonas arctica*	98	HG795046	Phenantrene

Bacterial strains isolated from Caleta Cordova were affiliated to genera already described as hydrocarbon-degrading. In particular, the isolates were referable to species adapted to cold, sub-Antarctic, environments. *Marinobacter antarcticus* showed cream-colored colonies, 1.5 mm diameter, halotolerant (optimum 3.0–4.0% NaCl w/v). The capability to degrade hydrocarbons was calculated as percentage of decrease in the amount of tetradecane compared to the un-inoculated control. The 30% of the added pollutant was degraded after 30 days. *Oleispira antarctica* is the first hydrocarbonoclastic bacterium identified in Antarctica [20], it grows with circular and cream-yellow colored colonies. After 30 days incubation, 40% of tetradecane was degraded. *Rhodococcus* was frequently described as hydrocarbon-degrading in single culture but also in consortia where it seems to improve the action of specialists hydrocarbonoclastic bacteria. Indeed, isolates belonging to *Rhodococcus* genus were less performant; after 30 days only about 15% of tetradecane was degraded. The degradation of phenanthrene at 4°C was assayed after 45 days of incubation. Both isolated strains showed a low growth performance and consequently also the degradation of the pollutant did not reach more than 5% of degradation.

4 DISCUSSION

Coastal marine environments are the most vulnerable regions to anthropic pressure, requiring advanced technologies for their proper knowledge and monitoring [21], [22]. Since the recent increase of oil spills (i.e. Exxon Valdez, Deep Horizon etc.), several studies have started to investigate the effects on marine biota, also including microbes, related to the presence of hydrocarbon contaminants [2], [10], [23], [24]. It is well known that a disturbance can modify microbial community composition [25] and, in turn, structural changes result in functional changes. To this regard, a reduction in microbial diversity was reported to affect general metabolic activity as well as the functional ability of the microbial community to degrade pollutants [26]. Our study addressed the response to oil contamination of the microbial community present in water and sediments, taking into account both microbial metabolism –

by extracellular enzyme activity measurements – and composition – by isolation of oil-degrading bacteria.

Microbes are recognized to play a key role in bioremediation processes; therefore, it is extremely important to study the relationships between microbial community structure and metabolic activity as a key step to elucidate how the microbial assemblage responds to oil pollution. On the other hand, the metabolic potential of oil-degrading communities by itself depends on the structure and diversity of microbial assemblage [8]. One possible limitation when assessing the ecological consequences of oil pollution is related to the fact that most of the studies of hydrocarbon-degrading microorganisms refer to laboratory experiments, involving the cultivation of bacteria; this makes understanding and prediction of *in situ* dynamics of microbial communities in response to oil discharge quite difficult [27] and explain the importance of combining culture with direct methods such as the biochemical ones to achieve a better comprehension of microbial dynamics.

4.1 Microbial metabolism

In both the study sites, LAP was the prevalent enzyme; proteolytic activity is widespread in the microbial community of temperate environments. The bacterial activity values measured in Milazzo Gulf ranged in the same order of magnitude as those reported for other pelagic Mediterranean waters. The obtained enzymatic data suggested that microbial metabolism was significantly depressed at the polluted stations, pointing out the negative effects of contaminants on organic matter degradation process. Similar findings were reported by a previous study [28], where oil pollution was reported to alter the structure and function of the whole microbial community. In this sense, functional diversity has been indicated as a key factor to assess the recovery of the microbial community after oil spill [29]. Nevertheless, controversial results on the effects of hydrocarbons on microbial metabolism are available in the scientific literature. In contrast with our results, Ziervogel et al. [30] detected, until 3 months after the Deepwater Horizon disaster, enhanced lipase activity and bacterial protein production in the surface (0–2 cm) sediment layers at the sites close to oil contamination, while leucine-aminopeptidase activity was depressed; this result indicated the stimulation of benthic microbial enzymatic hydrolysis of oil-derived organic matter that contained no proteins. A similar observation was reported by Ziervogel et al. [31] in a 21-day laboratory experiment in roller bottles to simulate microbial dynamics occurring in oil-contaminated waters; these Authors found bacterial-colonized oil aggregates, characterized by high lipase, LAP and glucosidase activities, although these last two enzymes are not directly related to primary oil-degradation. In deltaic lagoon sediments affected by persistent organic pollutants, Zoppini et al. [32] found increased prokaryotic production, enzymatic and community respiratory activities, pointing out the contribution of the benthic microbial community to sediment self-purification processes. Again, higher bacterial extracellular enzyme activities (lipase, α- and β-glucosidase, alkaline phosphatase, leucine aminopeptidase) and micro-aggregate counts were observed by Kamalanathan et al. [33] in a mesocosm study where tanks feed with Mexico Gulf water contaminated after the Deepwater Horizon oil spill, and added with chemical dispersants, were compared to controls without supplementation.

In the sediments of Milazzo and Caleta Cordova, enzyme activity rates up to 1–2 orders of magnitude higher than those recorded in the pelagic compartment were measured in our study, highlighting the role of benthic compartment as environmental archives and reservoirs of contaminants that can sink and accumulate within the sedimentary matrix.

WIT Transactions on Ecology and the Environment, Vol 242, © 2020 WIT Press
www.witpress.com, ISSN 1743-3541 (on-line)

4.2 Hydrocarbon-degrading bacteria

The isolation of bacterial hydrocarbon-degrading strains confirmed that the marine sites under study, known to have experienced oil contamination in the past, have a great potential in the response to oil-pollution. The growth of strains specifically adapted to use hydrocarbons and actively involved in the remediation processes let us hypothesize that they play a crucial role in the natural occurring clean-up processes also envisaging a biotechnological perspective for *in situ* bioremediation treatments. The results demonstrate the potential of both sites and evidence their peculiarities, with special attention to the temperature as one of the limiting factors for the success of bioremediation in marine environments.

In conclusion, in our study, both microbial structure and function in water and sediments appeared to be significantly affected by oil pollution. In order to minimize the environmental impact of oil spills and optimize the effectiveness of biodegradation measures, further estimates of the *in situ* metabolic potential of hydrocarbon-degrading bacteria and of the factors that limit microbially-catalyzed biodegradation will be needed.

Through long-term monitoring of polluted environments and the definition of baseline values related to the metabolic activity of the microbial community, it will be possible to assess the recovery of marine ecosystems after contamination.

ACKNOWLEDGEMENTS
This study was funded by the bilateral CNR and CONICET international cooperation research program "Global biodegradation network: ecological and molecular approach in the assessment of the oil pollution in the marine environment. A comparative and integrative study applied in two different coastal marine ecoregions (Mediterranean and Patagonian shelf)" (2009–2010). Particular thanks are due to Dr H. Dionisi (Laboratorio de Microbiología Ambiental, Centro para el Estudio de Sistemas Marinos, Centro Nacional Patagonico, CENPAT-CONICET, Puerto Madryn, Argentina), who supported the research activities in Patagonia.

REFERENCES
[1]   Yakimov, M.M., Timmis, K.N. & Golyshin, P.N., Obligate oil-degrading marine bacteria. *Current Opinions in Biotechnology*, **18**(3), pp. 257–266, 2007.
[2]   Sadooun, I.M.K., Impact of oil spills on marine life. *Emerging Pollutants in the Environment – Current and Further Implications*, eds M.L. Larramendy & S. Soloneski, IntechOpen: London, Chapter 4, 2015.
[3]   Cappello, S. et al., Microbial community dynamics during assays of harbour oil spill bioremediation: A microscale simulation study. *Journal of Applied Microbiology*, **102**(1), pp. 184–194, 2007.
[4]   Hassanshiahian, M., Emtiazi, G., Caruso G. & Cappello, S., Bioremediation (bioaugmentation/biostimulation) trials of oil polluted seawater: A mesocosm simulation study. *Marine Environmental Research*, **95**, pp. 28–38, 2014.
[5]   ITOPF.org, Environmental effects, 2016. www.itopf.com/knowledge-resources/documents-guides/environmental-effects/. Accessed on: 6 Aug. 2020.
[6]   Atlas, R.M. & Cerniglia, C.E., Bioremediation of petroleum pollutants. *Bioscience*, **45**(5), pp. 1–10, 1995.
[7]   Hazen, T.C., Prince, R.C. & Mahmoudi, N., Marine oil biodegradation. *Environmental Science and Technology*, **50**(5), pp. 2121–2129, 2016.

[8]   Head, I.M., Jones, D.M. & Röling, W.F., Marine microorganisms make a meal of oil. *Nature Reviews Microbiology,* **4**(3), pp. 173–182, 2006.

[9]   Howarth, R.W., Determining the ecological effects of oil pollution in marine ecosystems. *Ecotoxicology: Problems and Approaches. Springer Advanced Text in Life Sciences,* eds S.A. Levin, J.R. Kelly, M.A. Harwell & K.D. Kimball, Springer: New York, pp. 69–97, 1989.

[10] Nogales, B., Lanfranconi, M.P., Piña-Villalonga, J.M. & Bosch, R., Anthropogenic perturbations in marine microbial communities. *FEMS Microbiological Reviews*, **35**, pp. 275–298, 2011. DOI: 10.1111/j.1574-6976.2010.00248.x.

[11] Yakimov, M. et al., Natural microbial diversity in superficial sediments of Milazzo Harbor (Sicily) and community successions during microcosm enrichment with various hydrocarbons. *Environmental Microbiology*, 7(9), pp. 1426–1441, 2005.

[12] Commendatore, M.G. & Esteves, J.L., An assessment of oil pollution in the coastal zone of Patagonia, Argentina. *Environmental Management*, **40**, pp. 814–821, 2007.

[13] Crisafi, F., Giuliano, L., Yakimov, M.M., Azzaro, M. & Denaro, R., Isolation and degradation potential of a cold-adapted oil/PAH-degrading marine bacterial consortium from Kongsfjorden (Arctic region). *Rendiconti Lincei*, **27**(1), pp. 261–270, 2016.

[14] Winnepenninckx, B., Backeljau, T. & De Wachter, R., Extraction of high molecular weight DNA from molluscs. *Trends in Genetics*, **9**(12), p. 407, 1993.

[15] Lane, D.J., 16/23S rRNA sequencing. *Nucleic Acid Techniques in Bacterial Systematics*, eds E. Stackebrandt & M. Goodfellow, John Wiley: New York, pp. 115–175, 1991.

[16] Maidak, B.L., Olsen, G.J., Larsen, N., Overbeek, R., McCaughey, M.J. & Woese, C.R., The RDP (ribosomal database project). *Nucleic Acids Research*, **25**(1), pp. 109–111, 1997.

[17] Yakimov, M.M. et al., Phylogenetic survey of metabolically active microbial communities associated with the deep-sea coral *Lophelia pertusa* from the Apulian Plateau, Central Mediterranean Sea. *Deep-Sea Research Part I*, **53**(1), pp. 62–75, 2006.

[18] Rocchetti, L., Beolchini, F., Ciani, M. & Dell'Anno, A., Improvement of bioremediation performance for the degradation of petroleum hydrocarbons in contaminated sediments. *Applied and Environmental Soil Science*, **2**, pp. 1–8, 2011.

[19] Liu, C. & Shao, Z., *Alcanivorax dieselolei* sp. nov., a novel alkane-degrading bacterium isolated from sea water and deep-sea sediment. *International Journal of Systematic and Evolutionary Microbiology*, **55**, pp. 1181–1186, 2005.

[20] Yakimov, M.M. et al., *Oleispira antarctica* gen. nov., sp. nov., a new hydrocarbonoclastic marine bacterium, isolated from an Antarctic coastal seawater. *International Journal of Systematic and Evolutionary Microbiology*, **53**(3), pp. 779–785, 2003.

[21] Zappalà, G., Caruso, G. & Crisafi, E., The "SAM" integrated system for coastal monitoring. *Environmental Studies*, **8**, pp. 341–350, 2002.

[22] Zappalà, G., Caruso, G. & Crisafi, E., Coastal pollution monitoring by an automatic multisampler coupled with a fluorescent antibody assay. *Environmental Studies*, **10**, pp. 125–133, 2004.

[23] Caruso, G. et al., Microbial assemblages for environmental quality assessment: Knowledge, gaps and usefulness in the European Marine Strategy Framework Directive. *Critical Reviews in Microbiology*, **42**(6), pp. 883–904, 2016.

[24] Kimes, N.E., Callaghan, A.V., Suflita, J.M. & Morris, P.J., Microbial transformation of the Deepwater Horizon oil spill: Past, present, and future perspectives. *Frontiers in Microbiology*, **5**, p. 603, 2014.

[25] Morris, L., O'Brien, A., Natera, S.H.A., Lutz, A., Roessner, U. & Long, S.M., Structural and functional measures of marine microbial communities: An experiment to assess implications for oil spill management. *Marine Pollution Bulletin*, **131**, pp. 525–529, 2018.

[26] Delgado-Baquerizo, M. et al., Lack of functional redundancy in the relationship between microbial diversity and ecosystem functioning. *Journal of Ecology*, **104**(4), pp. 936–946, 2016.

[27] Prosser, J.I. et al., The role of ecological theory in microbial ecology. *Nature Reviews Microbiology*, **5**, pp. 384–392, 2007.

[28] Ortmann, A.C., Anders, J., Shelton, N., Gong, L., Moss. A.G. & Condon, R.H., Dispersed oil disrupts microbial pathways in pelagic food webs. *PLoS One*, **7**(7), pp. 1–9, 2012.

[29] Lee, H. et al., Importance of functional diversity in assessing the recovery of the microbial community after the Hebei Spirit oil spill in Korea. *Environment International*, **128**, pp. 89–94, 2019.

[30] Ziervogel, K., Joye, S.B. & Arnosti, C., Microbial enzymatic activity and secondary production in sediments affected by the sedimentation pulse following the Deepwater Horizon oil spill. *Deep Sea Research II*, **129**, pp. 241–248, 2016.

[31] Ziervogel, K. et al., Microbial activities and dissolved organic matter dynamics in oil-contaminated surface seawater from the Deepwater Horizon oil spill site. *PLoS One*, **7**(4), e34816, 2012.

[32] Zoppini, A. et al., Bacterial diversity and microbial functional responses to organic matter composition and persistent organic pollutants in deltaic lagoon sediments. *Estuarine, Coastal and Shelf Science*, **233**, 106508, 2020.

[33] Kamalanathan, M. et al., Extracellular enzyme activity profile in a chemically enhanced water accommodated fraction of surrogate oil: Toward understanding microbial activities after the Deepwater Horizon oil spill. *Frontiers in Microbiology*, **9**, 798, 2018.

METHODOLOGICAL APPROACH FOR THE DELIMITATION OF "NO BATHING AREAS" IN MARINE COASTAL ZONES CLOSE TO THE OUTLETS OF NATURAL AND ARTIFICIAL WATER COURSES IN BASILICATA REGION, ITALY

MICHELE GRECO[1,2*], NICOLA UNGARO[3], GAETANO CARICATO[3], GIOVANNI MARTINO[1],
LORENA DI GIUSEPPE[2], GIUSEPPE GIMBATTI[2], PATRIZIA MAURO[2], ANNA MONTELLA[2],
GIOVANNI MUSSUTO[2], MARIA E. SALERA[2], DOMENICO FARAONE[3],
CARMELA DI GRAZIA[3] & PASQUALE DE LUISE[2]
[1]Engineering School of University of Basilicata, Italy
[2]Regional Environmental Research Foundation – FARBAS, Italy
[3]Environmental Protection Agency of Basilicata, Italy

ABSTRACT

The Basilicata coasts, both Tyrrhenian and Ionian, are characterised by a considerable number of water courses both natural and artificial, which represent critical points for influence on marine bathing water quality. In such a context, according to the Water Framework Directive (2000/60/EC) and Water Bathing Directive (2006/7/EC), it has been necessary to carry out a systematic water quality check integrating the ordinary monitoring activity lead by the Regional Environmental Protection Agency of Basilicata (ARPAB) in order to provide a preliminary characterization of the marine waters for bathing purposes. The adopted methodological approach suggests to apply a 1-D model for pollutant transport and diffusion/dispersion in water bodies in order to perform a preliminary assessment of the width of the "no bathing area," including the water-course mouth. Calibration has been provided using episodic data collected in some of the artificial and natural water courses flowing into the Basilicata seas. Moreover, the assessment of the longshore width of these "no bathing areas" is generally aligned with the middle axis of the water course mouths. It has been obtained by simulating the diffusion/dispersion plume using episodic data and systematically collected ones during the bathing seasons, starting from 2011. In such a framework, the model results suggest a differentiation between Ionian and Tyrrhenian coasts due to the local littoral morphology as well as to the natural deltas and artificial outlets. Therefore, in "no bathing areas" of the Ionian coast, the methodological approach proposes a width of 40 m for artificial water course outlets and 100 m for natural water course outlets. However, the proposed "no bathing area" of the Tyrrhenian coast is 20 m wide with only one artificial water course outlet.

Keywords: no bathing area, water framework directive, bathing water directive, marine waters monitoring, diffusion model.

1 INTRODUCTION

The main goal of the European Water Framework Directive (WFD) issued by the European Commission [1] is to ensure water quality protection and improvement throughout the European Union (EU) by enforcing a common water resource management and control policy as well as by defining powerful strategies aimed to reduce sea water pollution and coastal resource overexploitation. A Directive was promulgated in 2000 (Dir 2000/60/EC), it defines the deadlines to reach the objective of a good ecological level of integrity for all the types of water bodies: rivers, lakes, transitional water, coastal water and groundwater.

---

* ORCID: http://orcid.org/0000-0002-3986-7117

WIT Transactions on Ecology and the Environment, Vol 242, © 2020 WIT Press
www.witpress.com, ISSN 1743-3541 (on-line)
doi:10.2495/WP200091

In Europe, bathing water quality management is regulated by the Directive 2006/7/EC (WBD) of the European Parliament and of the Council of 15 February 2006 [2], repealing the Directive 76/160/EEC [3]. The Directive 2006/7/EC aims to preserve, to protect and to improve the quality of the environment as well as to protect human health by supplementing the Directive 2000/60/EC. The Directive 2006/7/EC lays down provisions on monitoring and classification of bathing water quality, its management and its public information.

Water is a limited natural resource and-its quality should be protected, defended, managed and treated. In particular surface waters are renewable resources with limited capacity for recovery after a negative impact caused by human activities. In order to increase the efficiency and rational use of natural resources, the Directive 2006/7/EC should be closely coordinated with other Community legislation concerning water, such as the Council Directives 91/271/EEC [4], of 21 May 1991 on urban waste water treatment, and the Directive 91/676/EEC [5] of 12 December 1991 on the protection of water against pollution by nitrates from agricultural sources and the Directive 2000/60/EC of the European Parliament and of the Council, of 23 October 2000 establishing a framework for Community action in the field of water resources.

"Bathing water" is surface, current, or lake fresh water and sea water where bathing is expressly authorised or not prohibited. In Basilicata the monitoring of bathing waters takes place from 1 April to 30 September of each year, with monthly sampling frequency and the reference indicators of microbiological pollution are *Intestinal Enterococci* (I.E.) and *Escherichia coli* (E.C.). The concentration in the surface waters of these indicators highly depends on the efficiency of waste water collection and treatment systems as well as on the self-purification ability of the receiving water bodies. This ability is often favoured by natural factors such as the hydrological characteristics [6], [7] and the presence of vegetation [8], [9]. Surveying and quantifying microbial pathogens can be quite expensive and complex, so the assessment of water microbiological quality is based on the quantification of indicators of faecal pollution [10]–[12].

All along the Basilicata coasts (or Lucanian coasts), the only "no-bathing" areas are actually close to the deltas/outlets of water courses; it was stated by a regional decision as a precautionary principle without any scientific base. In such context, according to the WFD and WBD, in 2017 the Regional Environmental Research Foundation (FARBAS) proposed to carry out a systematic water quality survey in order to provide a preliminary characterization of the deltas/outlets sea waters quality status for bathing purposes. If not significant changes in the marine-coastal environment occur and the boundary variables (i.e. solar radiation, salinity, temperature, pH, and nutrient availability) are almost stable, the assessment of the E.C. concentration can be obtained performing only specific microbiological analyses [13], [14]. Among fecal coliforms, E.C. is recognized the best indicator of fecal pollution [15]. Furthermore, numerical forecasting modelisation is a robust tool capable to support monitoring activities outlining the criticalities of the physical coastal system. It also provides useful information related to the inland pressure scenarios and pollutant sources. From a modelling point of view, indeed, the integrated and combined use of one-dimensional models with a complex two or three-dimensional models might be a valuable facility for investigating the pollutant propagation in space and time of E.C. *in the coastal marine area* [16].

The aim of this study is to propose an expeditive methodological approach physically based on the searching of "no bathing areas" in order to reduce the pollution impact on the littoral and its effect on the human health and tourism. Moreover, this study improves the knowledge of the coastal water quality state.

2 THE LUCANIAN TYRRHENIAN AND IONIAN COASTS

The study concerns the whole Lucanian coast both on Ionian and Tyrrhenian littorals, interesting about 37 km and 25 km respectively, paying particular attention to the outlets of the natural and artificial water courses.

The Ionian coast starts from Metaponto (NE) (Fig. 1) to Nova Siri (SW) where the rivers Bradano, Basento, Sinni, Cavone and Agri flow. The Tyrrhenian coast mainly concerns Maratea area.

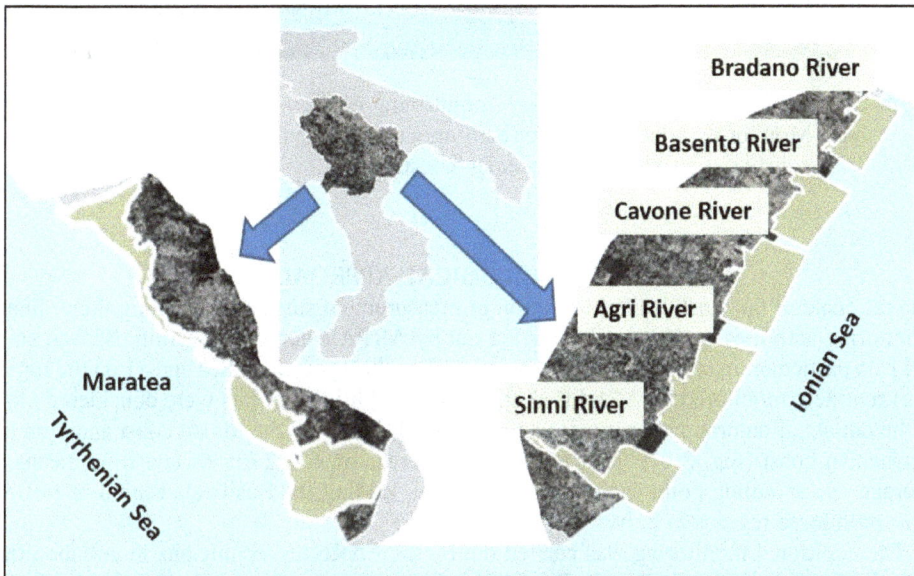

Figure 1:   Study areas along the Ionian and Tyrrhenian Lucanian coasts and the Site of Community Importance by Habitats Directive (92/43/EEC).

The Ionian littoral is characterized by the presence of Sites of Community Importance (SCI) *sensu* "Habitat Directive" (1992/43/CEE):

- SCI "Costa Ionica foce Bradano
- SCI "Costa Ionica foce Basento"
- SCI "Costa Ionica foce Cavone"

SCI *"Costa Ionica foce Agri"* The Tyrrhenian coast is characterized by the presence of Marine Nature Reserve *"Costa di Maratea"* rich in marine benthic communities and *Posidonia oceanica* meadows, as well as the following SCI:

- SCI "Acquafredda di Maratea"
- SCI "Marina di Castrocucco"
- SCI "Isola di S. Ianni e Costa prospiciente"

Moreover, along the Ionian coast there are 9 outlets of the artificial remediation channels which drains water to improve agricultural activities, which represents the main economic source of the area.

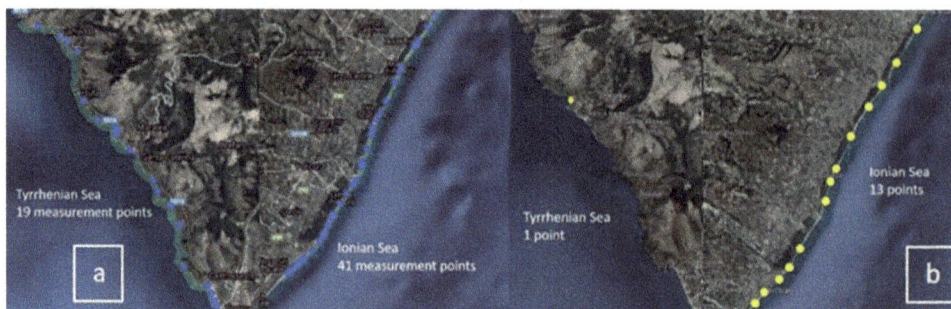

Figure 2:   (a) Bathing water, ordinary monitoring network of ARPAB [17]; and (b) Monitoring added points for bathing waters close to natural and artificial water course outlets along the Lucanian coasts.

3 METHODOLOGICAL APPROACH

The rationale of the work is the realization of measurement surveys integrating the ordinary monitoring activities systematically carried out by ARPAB during the bathing season since 2011. In particular, in addition to the 60 points (19 on the Tyrrhenian Sea and 41 on the Ionian Sea) routinely monitored by ARPAB (Fig. 2(a)), 14 additional points were considered close to the outlets of natural and artificial water courses [17], 13 on the Ionian coast and 1 on the Tyrrhenian coast (Fig. 2(b)). That is, 74 control areas on 61.52 km of coastline means in average one sampling point each 830 m of littoral, leading the Basilicata region one of the most monitored regions in Italy.

The additional monitoring was carried out by the FARBAS technicians in collaboration with the personnel of the ARPAB. Sampling and analyses were carried out monthly during the bathing season (April–September) for three-year period (2017–2019).

In the monitoring surveys, a multi-parameter YSI Pro Plus probe was used for the additional parameter pH, ORP, conductivity, dissolved oxygen and pressure (as proxy of depth). The microbiological analyses [18] of the samples were carried out at the ARPAB Microbiology Laboratory. The methods employed were the UNI EN ISO 7899-2 [19] for *Intestinal Enterococci* (I.E.) and UNI EN ISO 9308-1 [20] *Escherichia coli* (E.C.) respectively [21].

Each added monitoring point was fixed aligned to the middle of the outlet, both for natural and artificial water courses, while the width of the investigated area for the study was estimated through the implementation of 1-D model coherent to the microbiological indicators E.C. and I.E. temporal and spatial diffusion dynamics.

Further, the 1-D exponential spatial decaying law was adopted as first order assessment of the width of the no-bathing area to be considered in the study, while a 2-D model Delft3D [22], [23] software was subsequently applied in order to obtain the analysis scenarios for further evaluation of the "no bathing area" related to the local coastal hydrodynamics and health risk according to the Directive 2006/7/EC.

Through the Delft3D-Flow-Wave-D-WaQ coupled modules [24], [25], critical scenarios for hydrodynamic predictive simulations with limit values of E.C. and I.E. for marine waters (E.C. 500 UFC/100ml and I.E. 200 UFC/100ml) were assumed in order to evaluate the distance from the outlets where there is not microbiological pollution risk.

Specifically, the Delft3D-D-WaQ (Water Quality) module exploits the hydrodynamic conditions (velocity, water elevation, density, salinity, viscosity and diffusivity) calculated in the module Delft3D-Flow-Wave for given climate sea wave condition (wave height, wavelength and wave period) and simulates water quality by solving the diffusion equation on a fixed computational grid for a wide range of modelling substances.

The simulation scenarios refer to well defined space-time domain and time-spatial conditions of short, medium or long period that can be simulated on more or less detailed spatial domains depending on the required performances and available data [26], [27]. Therefore, the choice of the model and the subsequent methodological approach is a crucial point, since systematic and/or numerical errors can be generated compromising the value of the results as consequence of the approximation induced by the quality of the employed data [28]. The Delft3D-Flow-Wave-D-WaQ coupling module, both in steady and unsteady conditions, is implemented and calibrated on pre-processing, processing and post-processing operations. The pre-processing phase deals with the preliminary analysis and characterization of the topo-bathymetric and climate wave input data, while the processing phase runs the simulation scenarios on a computational grid under defined initial and boundary conditions (I.E. and E.C. initial concentrations at the outlets) (Fig. 3).

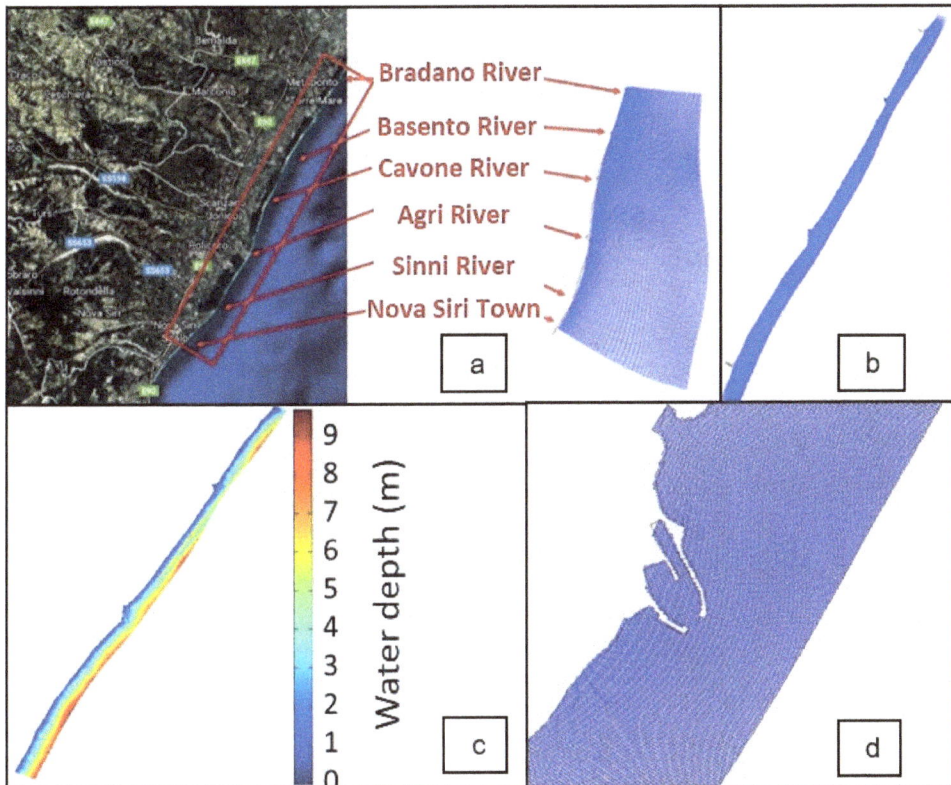

Figure 3: (a) Computational domain; (b) Computational grid; (c) Bathymetric domain; and (d) Detail of computational grid around the port (Argonauti harbour and Basento River delta).

4 RESULTS AND DISCUSSION

The proposed methodological approach was applied for the assessment of the "no bathing area" to be assumed for further study on the marine water quality. The preliminary implementation of 1-D diffusion model refers to episodic data collected on one of the nine artificial channel outlets (Nova Siri) in September 2016. In such a case, the measurements have been collected in the middle of the mouth and at 50 m and 150 m far away from the outlet along the coast, both sides, following ordinary procedures with few centimetres of wave height and calm weather conditions. Fig. 4, in the limits of the limited amount of data, reports the calibration curves for both parameters E.C. and I.E. and the relative decaying curves which allow us to assume the width of the study areas across the artificial water course outlet of about 40 mt. It is useful outline that the highest observed value for both parameters, 2600 and 634 UFC/100ml for E.C. and I.E. respectively, sensitively exceed the regulation values admitted for bathing into fresh waters but the rate of decay is reasonable high in the middle of outlet (BF curves: E.C.1000 UFC/100ml and I.E. 500 UFC/100ml threshold values for fresh-internal waters).

Moreover, the 2-D model Delft3D-Flow-Wave-D-WaQ coupling module was applied to the whole of the Ionian coast in order to have further information concerning the response of the system and the robustness of the preliminary assessment. In detail, the simulations were carried out assuming the following inputs and generally conservative boundary conditions: almost quiet sea water condition (5 cm wave height), no wind and longshore currents (typical of the seasonal climate features) which generally reduce the pollutant dispersion effect [29], average summer water discharges for both rivers and artificial channels and the values of I.E. and E.C. in such discharges were fixed to 500 and 1000 UFC/100ml respectively, corresponding to the limits of Law adopted for bathing compliance in the fresh water bodies.

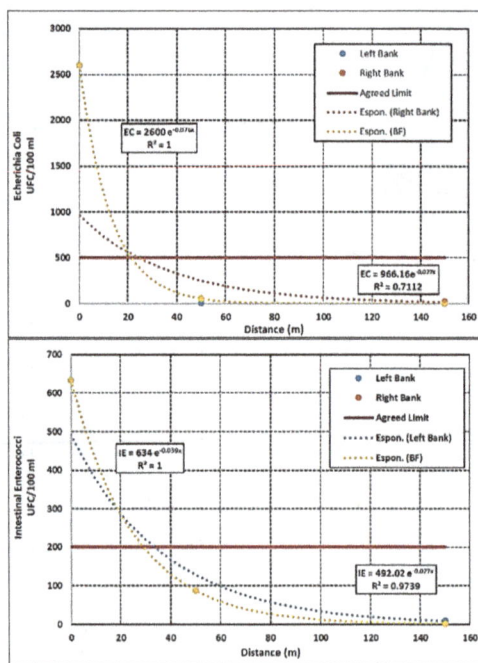

Figure 4:   Outputs of the 1-D model coherent to E.C. and I.E. temporal and spatial diffusion dynamics to assess the width of the study areas across the water courses outlets.

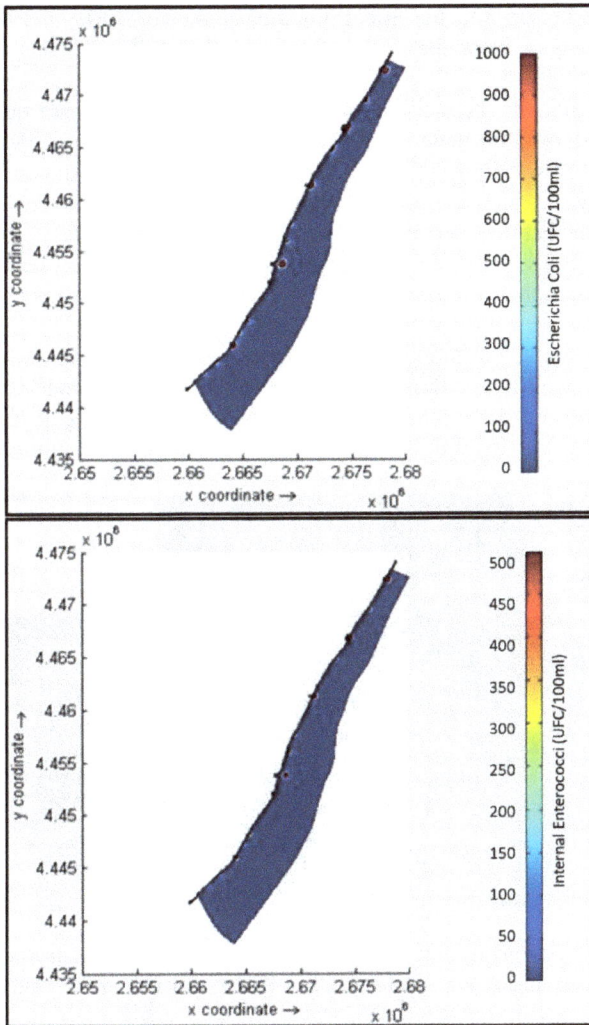

Figure 5:   Study scenario according to the input values of 1000 CFU/100 ml for E.C. and 500 CFU/100 ml for E.I. at the water course outlets.

Fig. 5 reports the results of the 2-D model Delft3D-Flow-Wave-D-WaQ coupling module simulations referred to the whole littoral for both parameters, in which are shown the cumulative effects along the coast. Moreover, the simulation where addressed to define the width of the "no bathing area" to be considered in order to carry out monitoring activities for the outlets influence assessment.

In Figs 6 and 7 are reported, as examples, the numerical results found for natural and artificial outlets, respectively, and the evaluated distance at which the concentration of the microbiological indicators are lower than the regulation limits for marine waters. In details, Fig. 6, which refers to two different deltas (Cavone River and Basento River) outlining the numerical results under the imposed boundary conditions previously discussed. That is, it shows that close to the river deltas the values of concentration fall within the limits of Law for the bathing water for both the microbiological indicators (I.E. and E.C.) already at a distance of about 30 m from the middle of the outlet well according to the expeditive 1-D model forecasting.

On the other hand, for the artificial water courses the outputs are quite different with respect to those observed for natural rivers even due to the low average values of the water discharge assumed flowing into the sea. That is, the discharges assumed in the 2-D model Delft3D-Flow-Wave-D-WaQ coupling module simulations are the average seasonal values observed during the bathing period (April–September), more or less close to 0.1 cumes, which allow us to have a mean rates for both microbiological pollutant sensitively low. Thus, the numerical simulations of artificial water course outlets report a negligible width of the no bathing area.

In such a framework, the model results suggest a differentiation between natural and artificial outlets as well as for the Ionian and Tyrrhenian coasts, the last due to the local littoral morphology and also general climate wave which induce further mixing effect influencing the decay curves of both E.C. and I.E. microbiological pollutant concentrations.

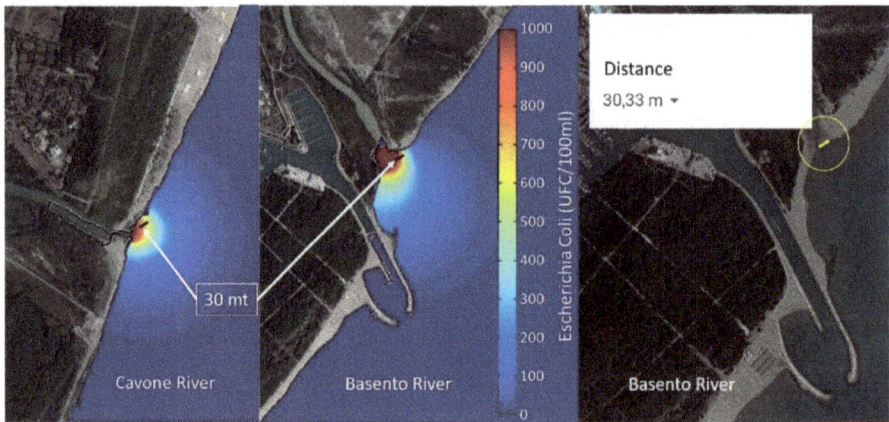

Figure 6:   Simulation results for the river deltas under the boundary conditions of 1000 CFU/100 ml for E.C. and 500 CFU/100 ml for E.I. at the water course outlets.

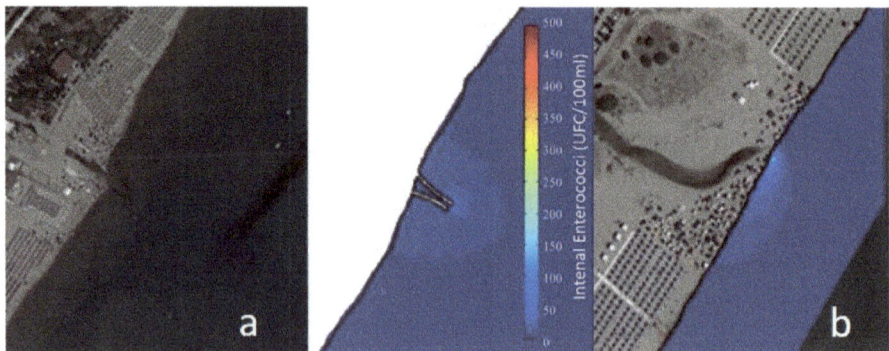

Figure 7:   Simulation results for the artificial outlets under the boundary conditions of 1000 CFU/100 ml for E.C. and 500 CFU/100 ml for E.I. at the water course outlets.

5 CONCLUSIONS

The routine monitoring plan for Basilicata's bathing waters provides to check 60 bathing areas on about 61.5 km of littoral.

Starting from the 2017 bathing season, additional control points were added in marine-coastal areas corresponding to the water courses outlets to evaluate the load of microbiological pollutants coming from the rivers and canals. The proposed methodological approach physically based was applied for the delimitation of "no bathing areas" in order to minimize the health risk.

Under the assumed boundary conditions of 1000 and 500 UFC/100 ml for E.C. and I.E. respectively, the final results give a precautional value of the width of "no bathing areas" to be 40 m for artificial outlets and 100 m for natural deltas along the Ionian coast, while the value for the artificial outlet has been fixed at 20 m on the Tyrrhenian coast (Fig. 8).

Finally, the study provides the scientific support for robust a suitable identification and delimitation of the "no bathing area" along the coasts even related to those marine-coastal zones potentially involved in short-term pollution events or anomalous circumstances, according to the Directive 2006/7/EC.

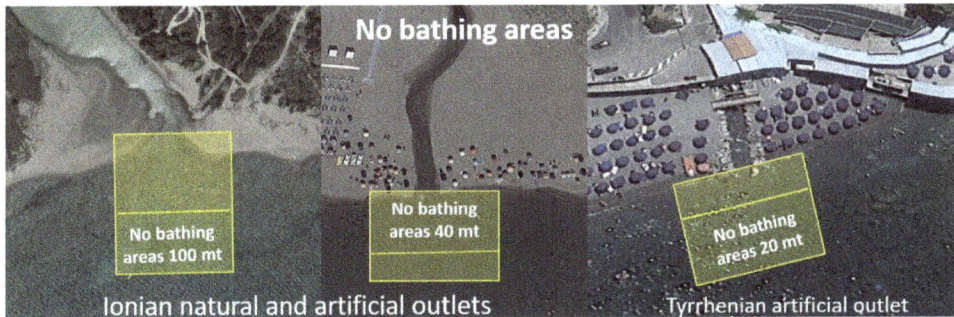

Figure 8: No bathing area for natural and artificial outlets on Ionian and Tyrrhenian seas.

ACKNOWLEDGEMENTS

This work has been granted by Regione Basilicata – DGR 1490/2014 – Project "Risk Communication and Social Mediation" and carried out by the Engineering School of University of Basilicata – SI-UNIBAS, Regional Environmental Research Foundation – FARBAS in collaboration with Environmental Protection Agency of Basilicata – ARPAB.

6 REFERENCES

[1]   EC-WFD, Directive 2000/60/EC of the European Parliament and of the Council of 23 Oct. 2000 establishing a framework for Community action in the field of water policy. European Water Framework Directive. Official Journal of the European Communities L 327/1, 2000.

[2]   Directive 2006/7/EC of the European parliament and of the council of 15 Feb. 2006 concerning the management of bathing water quality and repealing directive 76/160/EEC. Official Journal of the European Union L 64/37.

[3]   Council Directive of 8 Dec. 1975 concerning the Quality of Bathing Water (76/160/EEC)

[4]   Council Directive 91/271/EEC of 21 May 1991 concerning urban waste-water treatment

[5]   Council Directive 91/676/EEC of 12 Dec. 1991 concerning the protection of waters against pollution caused by nitrates from agricultural sources.

[6]   Malcangio, D. & Mossa, M., A laboratory investigation into the influence of a rigid vegetation on the evolution of a round turbulent jet discharged within a cross flow. *Journal of Environmental Management*, **173**, pp. 105–120, 2016. DOI: 10.1016/j.jenvman.2016.02.044.

[7]   Ben Meftah, M., Malcangio, D., De Serio, F. & Mossa, M., Vertical dense jet in flowing current. *Environmental Fluid Mechanics*, **18**, pp. 75–96, 2018. DOI: 10.1007/s10652-017-9515-2.

[8]   Ben Meftah, M., De Serio, F., Malcangio, D. & Mossa, M., Resistance and boundary shear in a partly obstructed channel flow. *River Flow 2016*, eds G. Constantinescu, M. Garcia & D. Hanes, Taylor & Francis Group: London, pp. 795–801, 2016. ISBN 978-1-138-02913-2.

[9]   Ben Meftah, M., Malcangio, D. & Mossa, M., Vegetation effects on vertical jet structures. *River Flow 2014*, eds A.J. Schleiss, G. De Cesare, M.J. Franca & M. Pfister, Taylor & Francis Group: London, 2014, pp. 581–588, 2014.

[10] Favero, M.S., Microbiological indicators of health risks associated with swimming. *American Journal of Public Health*, **75**, pp. 1051–1054, 1985. DOI: 10.2105/AJPH.75.9.1051.

[11] Noble, R.T., Moore, D.F., Leecaster, M., McGee, C.D. & Weisberg, S.B., Comparison of total coliform, faecal coliform, and enterococcus bacterial indicator response for ocean recreational water quality testing. *Water Research,* **37**, pp. 1637–1643, 2003. DOI: 10.1016/S0043-1354(02)00496-7.

[12] Cheung, W.H.S., Chang, K.C.K. & Hung, R.P.S., Variations in microbial indicator densities in beach waters and health-related assessment of bathing water quality. *Epidemiology and Infection,* **106**, pp. 329–344, 1991. DOI: 10.1017/S0950268800048482.

[13] Palazón, A., López, I., Aragonés, L., Villacampa, Y. & Navarro-González, F.J., Modelling of Escherichia coli concentrations in bathing water at microtidal coasts. *Science of The Total Environment*, **593–594**, pp. 173–181, 2017. ISSN 0048-9697. DOI: 10.1016/j.scitotenv.2017.03.161.

[14] Federigi I., Verani, M. & Carducci A., Sources of bathing water pollution in northern Tuscany (Italy): Effects of meteorological variables. *Marine Pollution Bulletin*, **114**(2), pp. 843–848, 2017. DOI: 10.1016/j.marpolbul.2016.11.017.

[15] Odonkor, S.T. & Ampofo J.K., Escherichia coli as an indicator of bacteriological quality of water: An overview. *Microbiology Research*, **4**, pp. e2, 2013.

[16] Bonamano, S. et al., Modeling the dispersion of viable and total Escherichia coli cells in the artificial semi-enclosed bathing area of Santa Marinella (Latium, Italy). *Marine Pollution Bulletin,* **95**(1), pp. 141–54, 2015. DOI: 10.1016/j.marpolbul.2015.04.030.

[17] Portale acque di balneazione – Ministero della Salute. www.portaleacque.salute.gov.it/ PortaleAcquePubblico/.

[18] American Public Health Association. Standard Methods for the Examination of Water and Wastewater, Washington D.C., 1996.

[19] UNI EN ISO 7899-2 Water quality – Detection and enumeration of intestinal enterococci – Part 2: Membrane filtration method (ISO 7899-2:2000).

[20] EN ISO 9308-1 Water quality – Enumeration of Escherichia coli and coliform bacteria – Part 1: Membrane filtration method for waters with low bacterial background flora (ISO 9308-1:2014).

[21] OMS: Guidelines for safe recreational water environments, OMS Ginevra, 2003.

[22] Delft University of Technology. SWAN Cycle III Version 40.72ABCDE User Manual, Delft, The Netherlands, 2009.

[23] Delft3D-FLOW, Simulation of multi-dimensional hydrodynamic flows and transport phenomena, including sediments, User Manual version 3.15, 24 giugno, 2011,

[24] Deltares, 2011, Delft3D-Flow User Manual, Delft, The Netherlands.

[25] Deltares, 2011, D-Water Quality User Manual (Documentation of the input file), Delft, The Netherlands.

[26] Greco, M. & Martino, G., Modelling of coastal infrastructure and delta river interaction on ionic Lucanian littoral. *Procedia Engineering*, **70**, pp. 763–772, 2014.

[27] Greco, M. & Martino, G., Vulnerability assessment or preliminary flood risk mapping and management in coastal areas. *Natural Hazards*, **82**(1), pp. 7–26, 2016.

[28] Greco, M. et al., Integrated SDSS for Environmental Risk Analysis in Sustainable Coastal Area Planning. *Computational Science and Its Applications – ICCSA 2018. ICCSA 2018. Lecture Notes in Computer Science, vol 10964*, eds O. Gervasi et al., Springer: Cham, 2018. DOI: 1007/978-3-319-95174-4_51.

[29] Menéndez A. & Laciana C., Pollutant dispersion in water currents under wind action. *Journal of Hydraulic Research*, **44**(4), pp. 470–479, 2006.

SECTION 4
WASTEWATER AND
WATER REUSE

USING DEWATERED SLUDGE FROM A DRINKING WATER TREATMENT PLANT FOR PHOSPHORUS REMOVAL IN CONSTRUCTED WETLANDS

NURIA OLIVER[1], CARMEN HERNÁNDEZ-CRESPO[2], MARIA PEÑA[1], MIGUEL AÑÓ[1],
ADRÍAN MARTÍNEZ[3] & MIGUEL MARTÍN[2]
[1]Global Omnium, Spain
[2]Instituto De Ingeniería Del Agua Y Medio Ambiente, Universitat Politècnica De València, Spain
[3]Departamento Ingeniería Hidráulica Y Medio Ambiente, Universitat Politècnica De València, Spain

ABSTRACT

This study evaluates the applicability of drinking water sludge as an active substrate in vertical subsurface flow constructed wetlands (VFCW) for advanced wastewater treatment. This treatment enables the achievement of greater phosphorus removals, as well as other pollutants, reaching concentrations below the required limits and obtaining suitable water for different uses, particularly environmental ones. To this end, two prototypes, one of them operating intermittently and another one with continuous flow, have been installed at Quart-Benàger urban wastewater treatment plant (WWTP), facility property of Entidad Pública de Saneamiento de Aguas Residuales (EPSAR) of the Valencian region. Preliminary results indicate that the reuse of the drinking water sludge as an active substrate in VFCW improves the quality of WWTP effluents, minimizing their impact on the receiving aquatic environments.

Keywords: pilot scale, phosphorus removal, reuse, drinking water sludge, subsurface flow constructed wetlands.

1 INTRODUCTION

Despite the fact that in recent decades a great effort has been made to improve the environmental quality of European water bodies, only 40% of the surface waters reached the ecological status "good" or "very good" stipulated by the Water Framework Directive (WFD, Directive 2000/60/EC) during the surveillance monitoring period 2010–2015 [1]. These data show the need to look for sustainable and economically viable solutions that guarantee the supply of good quality water to these ecosystems, allowing them to improve their ecological and chemical status, as well as maintaining the multiple services that they provide us.

In addition, the UN [2] warns that the world population is expected to increase by more than 1 billion inhabitants in the next 10 years, so that by 2030 there will be 8.6 billion inhabitants. This fact is going to lead to an important increase both in the impacts and in the demands of natural resources, and more specifically on water. This problem is aggravated in arid or semi-arid regions where water scarcity is a chronic problem, and it is essential to carry out adequate comprehensive water management so that it is returned to the receiving environment with adequate quality. In this sense, a higher purification level guarantees a better quality of the water in the receiving environment, with better oxygenation levels, thus mitigating the effects of the increase in temperature associated with climate change [3].

In Spain, 72% of the municipalities have a population of less than 2,000 inhabitants, according to the National Statistics Institute, some of which do not have a wastewater treatment plant (WWTP) or it does not work properly. Together, these municipalities house a total population of two million seven hundred thousand inhabitants, which supposes a polluting load of approximately 162 tons of organic matter per day. Although the pollutant load of each of these municipalities may seem small, the problem lies in the fact that their discharges are normally located in headwaters of rivers or in small streams, with limited dilution capacity and in many cases of high ecological value. Therefore, in small towns, the water sector must adapt to the new regulatory requirements regarding water, applying technologies adapted to the rural environment, such as constructed wetlands (CW). In this sense, the WFD involved a paradigm shift in the control of discharges into the aquatic environment, going from having to meet discharge limits to meeting quality objectives in the water bodies, which implied that the discharge authorizations should contemplate these environmental objectives when setting the discharge requirements in each case.

CW are water treatment infrastructures where physical, chemical and biological processes normally occurring in natural wetlands are reproduced in a controlled way, giving a treated effluent. In recent decades, numerous case studies have shown that these systems are a real alternative to conventional wastewater treatment technologies in small populations (<2000 h.e.). Among its advantages, we can highlight: high efficiency in the removal of organic matter and suspended solids, very simple maintenance, no or little energy consumption, little generation of sludge and no use of chemical products. However, it is difficult to achieve a high elimination of inorganic nutrients (orthophosphates and ammonium) [4]. The most widely used technology for the removal of phosphorus in CW is based on the use of active substrates as filling material [5], [6].

Another environmental problem that humanity faces is the large amount of solid waste generated from productive activities and whose management is inefficient. In the field of the Integral Water Cycle, and specifically in the water purification process, large amounts of sludge are produced, as a result of the applied physical-chemical treatments. Currently, it is estimated that each year around 3.8 kg·person^{-1} of this "waste" is generated worldwide [7], and its global production probably exceeds 10,000 tn·day^{-1} [8], with expectations that it will increase in the future.

In Europe, this waste is considered non-hazardous waste on the European waste list (LER 190902) and therefore it is usually disposed in landfills, thereby increasing the overall costs, both economic and environmental, of water treatment [9]. This fact highlights the need to find efficient and sustainable alternatives for its management.

One of the properties that characterizes this sludge, when aluminium salts are used as a chemical coagulant during the water purification process, is that it has a high chemical affinity for phosphorus (P). Previous investigations have verified that its use in wastewater treatment improves the processes of adsorption and chemical precipitation of phosphorus [10]–[13] in addition to other contaminants.

In this work, the applicability of the sludge produced in one Drinking Water Treatment Plant (DWTP) as an active substrate in CW with vertical subsurface flow (VFCW) is evaluated. These wetlands would be destined for an advanced urban wastewater treatment, after a secondary treatment, with the aim of obtaining an enhanced effluent suitable for different uses, especially environmental (wetland maintenance, ecological flows, etc.). The initial hypothesis is that the sludge still maintains a certain capacity for the elimination of substances through sorption or chemical precipitation processes and that plants and biofilm can complement the treatment with other biochemical processes.

Underlying the spirit of the project is the objective of moving towards a sustainable development, applying the principles of circular economy in the Integral Water Cycle (Fig. 1). To evaluate the efficiency of this treatment, two VFCW prototypes have been installed in the Quart-Benàger WWTP (Valencia) for treating a part of the effluent from the secondary treatment (prior to disinfection).

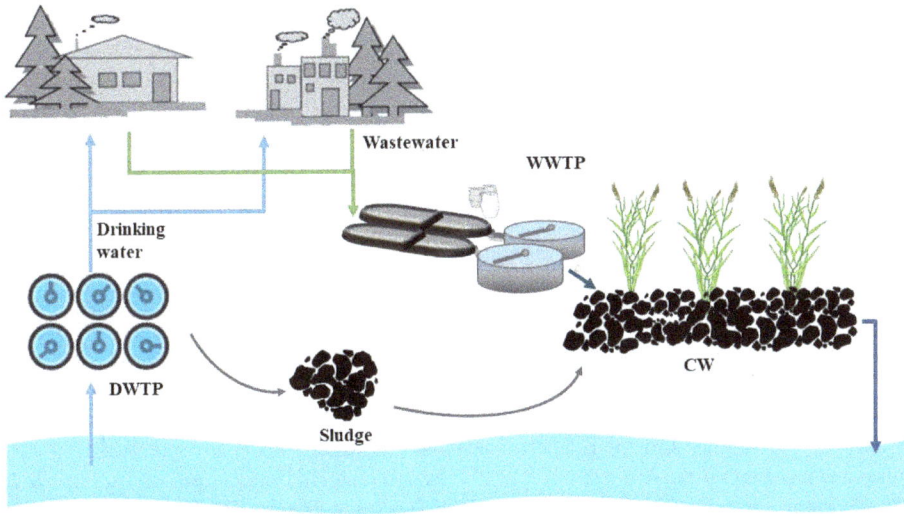

Figure 1:   Use of the sludge produced in the DWTP as an active substrate for CW in the final phase of the Integral Water Cycle [14].

2 MATERIALS AND METHODS

Two prototypes with a surface of 1 m^2 were installed in the Quart-Benàger WWTP, facility property of EPSAR, attached to Conselleria de Agricultura, Desarrollo Rural, Emergencia Climatic y Transición Ecológica, QB-WWTP hereafter (Valencia, Spain). They were filled with two layers of gravels (size: 10–11 mm) at bottom and surface, and three intermediate layers of sludge from the DWTP La Presa (Manises, Valencia, Spain), with the granulometry and thickness indicated in Fig. 2. The planted vegetation was *Phragmites australis* (common reed), one of the typical species of the humid zones and commonly used in CW [15], with an initial density of 10 plants·m^{-2}.

The La Presa DWTP produces approximately 12 m^3·day^{-1} of sludge with a humidity of around 75%, which contains the particles present in raw water before treatment: suspended solids, flocculated mineral and organic materials, metallic hydroxides (iron and manganese), and the residues of the coagulants added for treatment, i.e. aluminium polychloride (Al$_2$(OH)$_3$Cl) aided by a flocculant (PoliDADMAC), and powdered activated carbon. After several dehydration processes, centrifuges, and drying beds, the sludge cake reaches a dry matter content of 33%. Once dried, the cakes were grinded, obtaining the grain sizes indicated in Fig. 2. The aluminium content was 709 mg·g^{-1} (dry weight).

One of the prototypes worked intermittently (VFCW-1), through several equal cycles throughout the day (filling, contact time and draining), while the other prototype operated at continuous flow (VFCW-2), the substrate remaining always flooded (Fig. 2).

The study has been divided into two periods, attending to the different contact times tested in the HAFV-1 and covering the entire start-up of the prototypes. In HAFV-1 wetland, the contact time is defined as the time that elapses between the end of filling and the beginning of emptying. In HAFV-2, the contact time is the theoretical hydraulic retention time (porous volume/flow rate). The hydraulic load applied in each case, as well as the contact times maintained, are presented in Table 1.

The performance of both pilot VFCV was monitored as follows. Samples of the influent (secondary treatment effluent of QB-WWTP) and the effluent from each prototype were collected for subsequent physical-chemical analysis. The variables analysed were: electrical conductivity, pH and dissolved oxygen (DO), measured in situ, turbidity, total suspended solids (TSS) and volatile (VSS), COD (total), BOD_5 (total), total nitrogen (TN), ammonium ($N-NH_4^+$), nitrites ($N-NO_2^-$), nitrates ($N-NO_3^-$), total phosphorus (PT), phosphates ($P-PO_4^{3-}$), aluminium (Al) and *Escherichia coli* (E. Coli). The latter was chosen as indicator of faecal pollution considered in water reuse regulation [16], [17]. For the physical parameters and soluble compounds, the frequency of the determinations has been 3 times per week ($n = 45$), while for the total compounds and the pathogens it has been 1 time per week ($n = 16$). All the chemical analyses have been carried out following standardized methods [18].

Finally, a statistical analysis was carried out to compare the existence of significant differences between the concentrations measured in the influent and in the effluent of each prototype, as well as between both effluents. If the samples presented normal distribution, the Student's t-test was applied for related samples, otherwise, the Friedman test was applied to analyse the relationship of different variables simultaneously and the Wilcoxon test to compare between pairs. All statistical analyses were carried out with a confidence level of 95%, assuming an error of 5%, using SPSS 15.0 software for Windows (SPSS Inc. Chicago, USA).

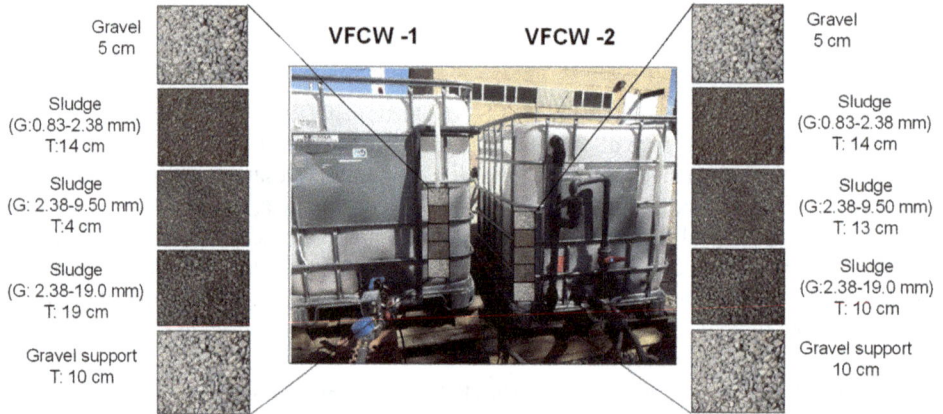

Figure 2: Prototypes image and grain size distribution (G: grain size; T: thickness).

Table 1: Hydraulic loading rates, cycles per day and contact times.

	Hydraulic loading rate (m³ m⁻² day⁻¹)		Contact time (minutes)	
	Period I	Period II	Period I	Period II
VFCW-1	3.75 ± 0.60 (22 cycles)	3.02 ± 0.39 (18 cycles)	30	45
VFCW-2	3.09 ± 0.48	3.00 ± 0.54	120	120

Table 2:   Influent concentration to the VFCW prototypes (average ± standard deviation for each period). The discharge limits by European and local authorities are: TSS (35 mg·L^{-1}) COD (125 mg·L^{-1}), BOD5 (25 mg·L^{-1}), TN (10 mg·L^{-1}), TP (0.6 mg·L^{-1}).

Variable	Units	Period I	Period II
TSS	mg TSS L^{-1}	6.9 ± 7.5	6.7 ± 1.4
COD	mg O$_2$ L^{-1}	30 ± 5	34 ± 4
BOD$_5$	mg O$_2$ L^{-1}	10 ± 3	10 ± 2
TP	mg P L^{-1}	0.31 ± 0.13	0.59 ± 0.16
PO$_4^{3-}$	mg P-PO$_4^{3-}$ L^{-1}	0.18 ± 0.26	0.32 ± 0.38
TN	mg N L^{-1}	7.69 ± 1.71	9.51 ± 0.94
NH$_4^+$	mg N-NH$_4^+$ L^{-1}	2.23 ± 1.53	4.49 ± 2.17
NO$_2^-$	mg N-NO$_2^-$ L^{-1}l	0.68 ± 0.44	0.90 ± 0.28
NO$_3^-$	mg N-NO$_3^-$ L^{-1}	4.02 ± 1.46	4.12 ± 0.72
E. Coli	log$_{10}$ CFU 100 mL^{-1}	4.9± 0.3	5.1 ± 0.4

3 RESULTS AND DISCUSSION

3.1 Influent characteristics

The results of physic-chemical characteristics of the influent to the VFCW prototypes during the study period are shown in Table 2. It should be highlighted that the results indicate very low concentrations, due to the fact that this is the effluent from the secondary treatment of the WWTP, as previously mentioned, which must meet not only the discharge requirements to a sensitive area (91/271/EEC Directive) but those established for the TP by the water authority (Jucar Hydrographic Confederation). Thus, the average value of TP was less than 0.6 mg P L^{-1} and average TN was 10 mg N L^{-1}, with nitrates being the major component.

3.2 Removal efficiencies

The average concentrations of effluent and removal efficiencies for each variable, period and prototype are indicated in Table 3 and Fig. 3 respectively. The mean effluent concentration of TP was significantly lower than that measured in the influent ($p < 0.05$) for both prototypes and well below the limit required for this WWTP. It is in the order of 0.1 mg P L^{-1}, a value recommended for the Albufera Lake restoration [19]. Regarding phosphates, the concentrations measured in the effluents were also significantly lower than those of the influent ($p < 0.05$), this reduction being mainly related to the adsorption and precipitation processes in the active substrate. Proof of this is the high efficiency achieved since the beginning of the prototypes' start-up, with vegetation and a biofilm not yet developed. The VFCW-2 prototype presented a significantly lower concentration, thus a higher efficiency, probably because it has a higher proportion of sludge with a smaller grain size (Fig. 2). In this sense, it is well known that the higher the specific surface area of the substrate, the greater its adsorption capacity [20]. Nevertheless, despite the differences found, the removal efficiencies achieved in both prototypes can be considered high (Fig. 3), exceeding 80% with the continuous operation mode (VFCW-2). In the case of VFCW-1, the TP removal efficiency improved with increasing the contact time (period II) and could also be related to a greater development of biofilm and vegetation.

Table 3: Effluent concentration from the VFCW prototypes (average ± standard deviation for each period).

Variable	Units	Period I		Period II	
		VFCW-1	VFCW-2	VFCW-1	VFCW-2
SST	mg TSS L^{-1}	2.5 ± 1.6	1.9 ± 1.2	2.5 ± 0.5	2.3 ± 0.5
COD	mg O_2 L^{-1}	21 ± 1	19 ± 2	19 ± 2	21 ± 2
BOD$_5$	mg O_2 L^{-1}	4 ± 2	4 ± 1	5 ± 2	5 ± 1
TP	mg P L^{-1}	0.12 ± 0.02	0.05 ± 0.01	0.18 ± 0.03	0.07 ± 0.02
PO$_4^{3-}$	mg P-PO$_4^{3-}$ L^{-1}	0.06 ± 0.04	0.01 ± 0.02	0.08 ± 0.03	0.03 ± 0.03
TN	mg N L^{-1}	7.92 ± 0.99	6.05 ± 1.26	9.19 ± 0.59	4.90 ± 0.78
NH$_4^+$	mg N-NH$_4^+$ L^{-1}	0.11 ± 0.13	1.89 ± 1.06	0.14 ± 0.15	2.63 ± 1.56
NO$_2^-$	mg N-NO$_2^-$ L^{-1}	0.09 ± 0.10	0.36 ± 0.24	0.09 ± 0.08	0.00 ± 0.01
NO$_3^-$	mg N-NO$_3^-$ L^{-1}	6.89 ± 1.47	3.28 ± 1.13	8.37 ± 0.87	1.58 ± 0.66
E. Coli	Log$_{10}$ CFU 100 m L^{-1}	4.2 ± 0.6	3.4 ± 0.4	3.8 ± 0.7	3.7 ± 0.5

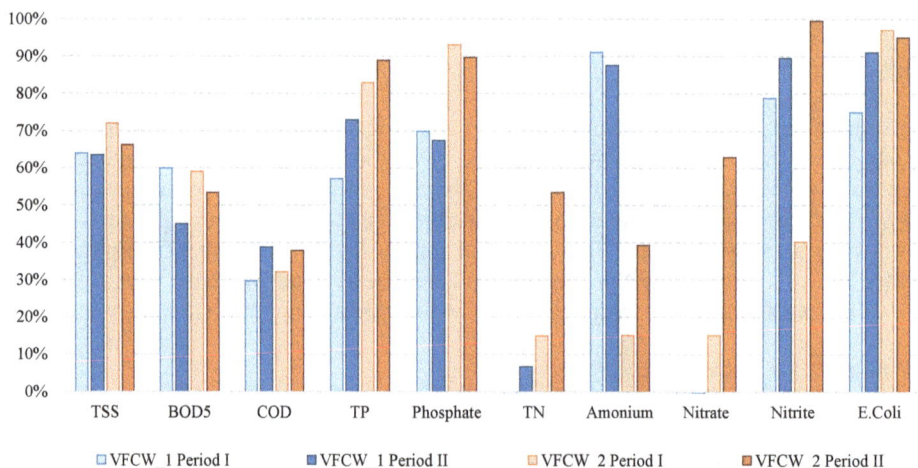

Figure 3: Mean removal efficiencies in each prototype and period.

Regarding TSS, both prototypes significantly reduced its concentration (p <0.05). In fact, the average value obtained, around 2 mg L^{-1}, is even lower than the average values measured in river water bodies protected by Red Natura 2000 [21]. Similarly to TP, the TSS retention efficiency was high from the beginning, greater than 60%. This TSS reduction is associated with the combination of sedimentation processes (as they slowly pass through the porous bed) and filtration.

The organic matter present in the influent to the prototypes, measured as COD and BOD$_5$, was low because it is the effluent from the secondary treatment of the QB-WWTP, thus can be characterized as slowly biodegradable. However, despite this, its concentration was further reduced as it passed through the VFCW, reaching mean concentrations of biodegradable organic matter below 5 mg $O_2 \cdot L^{-1}$. This value is below the limit established in the Spanish Hydrological Planning Instruction to define the good status of rivers (6 mg·L^{-1}) [22], therefore the effluent has a very good water quality. The average removal efficiencies achieved for COD were above 30% and up to 60% for BOD$_5$.

With regard to TN, the behaviour of both VFCW prototypes was different. VFCW-1 significantly reduced the ammonium concentration ($p<0.05$) because it has aerobic conditions thanks to the intermittent regime, which enables the oxygenation of the substrate during the draining stage, thus favouring the nitrification process. In line with this, TN did not decrease because ammonium was transformed into nitrates. Conversely, VFCW-2 has more anoxic conditions due to its continuous feeding, which promotes the denitrification process. Indeed, nitrates decreased significantly ($p<0.05$), thus reducing TN as well, whereas ammonium removal efficiency was low (15%) because of the anoxic conditions. Interestingly, this removal efficiency was significantly improved in the second period of the study (Fig. 3), coinciding with better environmental conditions for the nitrification-denitrification process (more developed root system and biofilm and higher temperatures), as well as a consumption of soluble inorganic nitrogen (preferably ammoniacal nitrogen) by vegetation. The latter, together with the increase of the contact time in HAFV-1, could have led to the net removal achieved in this prototype during the second period.

In addition to the parameters discussed so far, since it is intended to reuse the reclaimed water, it is crucial to guarantee very good quality conditions regarding the content of pathogenic microorganisms. Thus, the measured concentrations of E. Coli in the effluents of both prototypes (Table 2) were significantly lower than those measured in the influent ($p <0.05$), reaching average removal yields of up to 91% in HAFV-1 and up to 97% in HAFV-2. Moreover, in the case of HAFV-1, the removal efficiency improved significantly by increasing the contact time. Keeping in mind that one of the objectives is to meet the quality requirements for reusing the reclaimed water, this is an important finding in order to define the best operation conditions. It should be noted that during the second study period, both prototypes obtained a concentration of E. Coli in the effluent below 4.00 logarithmic units of CFU 100 mL^{-1}, thus accomplishing the limit value for the irrigation of woody crops, tree nurseries or industrial crops [16].

The removal of microorganisms is a highly complex process. Processes such as filtration, adsorption, and predation play a fundamental role in this type of wetland systems. In addition, the removal is dependent on the residence time and the granular medium, so that the smaller its diameter, the greater efficiency obtained.

Finally, given the presence of aluminium in the sludge, one might wonder if there is a risk of releasing this metal into the water flowing through the material. The results indicate that, concentrations never reached values higher than 0.2 mg·L^{-1}, which is the value stipulated by European Directive dealing with the sanitary criteria for the quality of water for human consumption [23].

4 CONCLUSIONS

The results of this study have shown that the sludge produced in "La Presa" DWTP can be used as main substrate of CW for improving the water quality of secondary effluents in WWTPs. It is a good adsorbent of phosphorus, as well as an appropriate media for the development of biofilm and typical vegetation of CW. Therefore, this technology has proven to be efficient for the advanced treatment of urban wastewater, significantly improving the quality of the effluents of WWTPs. More interestingly, the reclaimed water could be directly reused for some uses, without needing an additional disinfection treatment.

Thanks to the use of this kind of system, it is possible to achieve a phosphorus concentration in the effluent lower than 0.1 mg $P \cdot L^{-1}$, a value considerably lower than the limit established by the Jucar Basin Hydrological Plan (0.6 mg $P \cdot L^{-1}$), with HLR between 2.72–2.81 $m^3 m^{-2} d^1$. In addition, it contributes to reducing other pollutants such as total nitrogen, ammonium, BOD_5 and COD, still present, although in very low concentrations, in the secondary effluent of an activated sludge treatment.

The choice of one or the other type of mode of operation (HAFV-1 or HAFV-2) will depend on the quality objectives pursued for the effluent of the WWTP.

ACKNOWLEDGEMENTS

Research partially funded by the Generalitat Valenciana-IVACE (Valencian Institute of Business Competitiveness) and by the European Regional Development Fund (through the ERDF Operational Program for the Valencian Community 2014-2020).

The authors want to express their gratitude to the Public Entity for Wastewater Sanitation of the Valencian Community (EPSAR), as well as to the Cathedra "Aguas de Valencia-Universitat Politècnica de València," Department of Hydraulic Engineering and the Environment.

REFERENCES

[1] European Environment Agency (EEA). European waters—assessment of status and pressures. EEA Report No 7/2018. EEA, Copenhagen, 2018.

[2] UN. Department of Economic and Social Affairs, 2017. www.un.org/development/desa/es/news/population/world-population-prospects-2017.html.

[3] Vaughan, I.P. & Gotelli, N.J., Water quality improvements offset the climatic debt for stream macroinvertebrates over twenty years. *Nature Communications*, **10**, 2019. DOI: 10.1038/s41467-019-09736-3.

[4] Vymazal, J., Removal of nutrients in various types of constructed wetlands. *Science of the Total Environment*, **380**(1–3), pp. 48–65, 2007.

[5] Martín, M., Gargallo, S., Hernández-Crespo, C. & Oliver, N., Phosphorus and nitrogen removal from tertiary treated urban wastewaters by a vertical flow constructed wetland. *Ecological Engineering*, **61**, pp. 34–42, 2013.

[6] Vohla, C., Kõiv, M., Bavor, H.J., Chazarenc, F. & Mander, Ü., Filter materials for phosphorus removal from wastewater in treatment wetlands: A review. *Ecological Engineering*, **37**(1), 70–89, 2011.

[7] Zhao, Y., Liu, R., Awe, O.W., Yang, Y. & Shen, C., Acceptability of land application of alum-based water treatment residuals–an explicit and comprehensive review. *Chemical Engineering Journal*, **353**, pp. 717–726, 2018.

[8] Turner, T., Wheele, R., Stone, A. & Oliver, I., Potential alternative reuse pathways for water treatment residuals: Remaining barriers and questions: A review. *Water Air & Soil Pollution*, **230**, 2019.

[9] Hidalgo, A.M., Murcia, M.D., Gómez, M., Gómez, E., García-Izquierdo, C. & Solano, C., Possible uses for sludge from drinking water treatment plants. *Journal of Environmental Engineering*, **143**(3), pp. 04016088, 2017.

[10] Babatunde, A.O. & Zhao, Y.Q., Constructive approaches toward water treatment works sludge management: An international review of beneficial reuses. *Critical Reviews in Environmental Science and Technology*, **37**(2), pp. 129–64, 2007.

[11] Babatunde, A.O. & Zhao, Y.Q., Forms, patterns and extractability of phosphorus retained in alum sludge used as substrate in laboratory-scale constructed wetland systems. *Chemical Engineering Journal*, **152**(1), pp. 8–13, 2009.

[12]   Babatunde, A. O., Zhao, Y.Q. & Zhao, X.H., Alum sludge-based constructed wetland system for enhanced removal of P and OM from wastewater: Concept, design and performance analysis. *Bioresource Technology*, **101**(16), pp. 6576–79, 2010.

[13]   Zhao, Y.Q., Babatunde, A.O., Razali, M. & Harty, F., Use of dewatered alum sludge as a substrate in reed bed treatment systems for wastewater treatment. *Journal of Environmental Science and Health – Part A Toxic/Hazardous Substances and Environmental Engineering*, **43**(1), pp. 105–10, 2008.

[14]   Zhao, Y.Q., Babatunde, A.O., Hu, Y.S., Kumar, J.L.G. & Zhao, X.H., Pilot field-scale demonstration of a novel alum sludge-based constructed wetland system for enhanced wastewater treatment. *Process Biochemistry*, **46**(1), pp. 278–283, 2011.

[15]   Vymazal, J., Plants used in constructed wetlands with horizontal subsurface flow: A review. *Hydrobiologia*, **674**, pp. 133–156, 2011.

[16]   REAL DECRETO 1620/2007, de 7 de diciembre, por el que se establece el régimen jurídico de la reutilización de las aguas depuradas.

[17]   REGULATION (EU) 2020/741 of the European Parliament and of the Council of 25 May 2020 on minimum requirements for water reuse.

[18]   APHA. Standard Methods for the Examination of Water and Wastewater. 21st ed., American Public Health Association/American Water Works Association/Water Environment Federation, Washington DC, 2005.

[19]   Ministerio de Medio Ambiente, Estudio para el desarrollo sostenible de l'Albufera de Valencia, 2005. http://www3.chj.gob.es/albufera/01_WEB_ED/indexAV1.htm.

[20]   Yang, Y., Tomlinson, D., Kennedy, S. & Zhao, Y., Dewatered alum sludge: A potential adsorbent for phosphorus removal. *Water Science and Technology*, **54**(5), pp. 207–13, 2006.

[21]   Generalitat Valenciana. Evaluación de los Datos de Calidad de Aguas para el Seguimiento de Masas de Agua Fluvial en los Espacios Red Natura 2000. Informe técnico 04/2019, 2012. www.agroambient.gva.es/es/web/biodiversidad/evaluacion-de-los-datos-de-calidad-de-aguas-para-el seguimiento-de-masas-de-agua-fluvial-en-los-espacios-red-natura-2000.

[22]   ORDEN ARM/2656/2008, de 10 de septiembre, por la que se aprueba la instrucción de planificación hidrológica.

[23]   Directive 98/83/EC of 3 November 1998 on the quality of water intended for human consumption.

NATURE-BASED WATER TREATMENT SOLUTIONS AND THEIR SUCCESSFUL IMPLEMENTATION IN KATHMANDU VALLEY, NEPAL

ZUZANA BOUKALOVÁ[1,2], JAN TĚŠITEL[2,3] & BINOD DAS GURUNG[2,4]
[1]VODNÍ ZDROJE, a. s., Czech Republic
[2]METCENAS o. p. s., Czech Republic
[3]AMBIS a. s., Czech Republic
[4]Czech University of Life Sciences, Czech Republic

ABSTRACT

Surface and groundwater in cities and downstream urban areas may suffer from serious pollution from point and diffuse sources from upstream and in-catchment, which might have a negative impact on the ecology, quality of life and land values of the city. Enhanced nature-based treatment solutions (such as constructed wetlands) have the potential to remove pollutants from the water (e.g. storm water, urban run-off, river water and wastewater) that will lead to improved water quality and water use efficiency. Such natural treatment measures, when well organised and integrated into the overall urban planning and design, can also contribute to climate adaptation by reducing drought/flood risk and constitute attractive components of the urban landscape. The constructed wetlands are a strategic nature-based technology for Nepal, where discharge of untreated wastewater into rivers, lakes or any other water body is a common practice. Constructed wetlands are highly efficient in removing organic, insoluble substances and some pesticides, and their construction and operation are both simple and cost-effective. However, the main conditions of the successful implementation of the constructed wetlands in Nepal (and the other developing countries) should be clearly stated. Legislation and standards in Nepal are weak and, therefore, wastewater treatment is not a priority for city governments and private institutions. Despite constructed wetlands being a low-cost technology, it might be difficult to convince people to pay for their wastewater treatment. The key issue is "who" takes the responsibility for their maintenance and how this responsibility is defined and granted. Our paper discusses the best and worst practices in the Kathmandu Valley and the conditions that could influence the successful implementation of the nature-based treatment solutions in developing countries more generally.

Keywords: water management, nature-based water treatment solutions, constructed wetlands, pollution control, land management, communities, Nepal.

1 INTRODUCTION

Constructed wetlands are highly efficient in removing organic and insoluble substances and could be as well used for the degradation of pesticides, but this efficiency is highly variable for different substances. The greatest efficiency of wetlands has been demonstrated for organo-chlorinated pesticides, organo phosphates, Pyretroid and Strobin, on the contrary, the lowest for Triazins and uric acid derivatives. Pesticide degradation generally increases with the increasing value of the adsorption coefficient, but this dependence is not linear and too strong [1].

More, the construction and operation of the constructed wetlands are both simple and cost-effective, with low energy consumption (if any).

In Nepal, the most suitable location for the operation of constructed wetlands is the Kathmandu Valley where population is concentrated in three big cities (Kathmandu, Lalitpur and Bhaktapur Districts). At the same time, none of the five wastewater treatment plants, constructed in the past in the Kathmandu Valley is fully functional as of 2019 (an activated sludge plant at Guheshwori, non-aerated lagoons at Kodku and Dhobighat, and aerated lagoons at Sallaghari and Hanumanghat) [2].

Management of wastewater – particularly from households – is becoming an increasing problem in the Kathmandu Valley due to the increasing migration to cities [3]. Kathmandu Valley is characterized by high population growth (estimated to be 6.6% per annum) and high population density (estimated at more than 2,500 persons per km^2). The total population of Kathmandu Valley was estimated at 2.51 million in 2011 (CBS, census 2011) and will probably reach 3.26 million by 2021.

The existing wastewater network has not been maintained or expanded to serve the spreading urban areas and increased population. This has resulted in untreated sewage being discharged directly into local watercourses. The rivers have become open sewers presenting severe public health risks, in particular to the urban poor. Legislation and standards in Nepal are weak and therefore wastewater treatment is not a priority for city governments, private institutions and industries. However, the communities care.

In 2013, a comprehensive solution to the situation was proposed: The Kathmandu Valley Wastewater Management Project (KVWMP) of the government of Nepal, which, however, was still not completed in 2019 (in December 2019, testing of its part, the newly reconstructed Guheshwori Wastewater Treatment Plant, was launched). This project is aimed at improving wastewater services in Kathmandu Valley through extensive investment in rehabilitating and expanding the sewerage networks; modernizing and constructing wastewater treatment plants (WWTPs); and supporting operational and financial improvements and capacity building. The work includes the rehabilitation and construction of new WWTP at Kodku (Patan), Sallaghari (Bhaktapur), Dhobighat (Kathmandu) and Guyesheshowri (Kathmandu). All WWTPs designed under this project will be rehabilitated or constructed in the land area owned by government [4]. This should ensure their proper operation, continuous specialist's supervision as well as sufficient electric power supply.

In the Kathmandu Valley, constructed wetlands could then support the wastewater management out of reach of the WWTPs and provide additional wastewater treatment for a number of households, especially in areas, where water retention at landscape is needed, groundwater level is continuously declining and local hydrogeological situation is favourable for the water infiltration.

The idea of our research in Kathmandu Valley and as well of this paper is to review the recent situation of the constructed wetlands in the area. Next, find the circumstances, that are influencing wetlands functioning and define the important issues for their successful maintenance. The part of our research work is as well the construction of the pilot wetland, which would serve as the good example in the area. Via this pilot plant we would like to demonstrate how the wastewater management problems could be solved by the nature-based solution favourable for both, the local community/owner of the area and the environment.

An ideal area for this pilot plant (constructed wetlands) here is e. g. Lalitpur where the water supply situation has been critical since 2006, when continuous decline in groundwater levels began due to a combination of causes such as large engineering operations, collapse of the local water supply network, increasing migration as well as excessive groundwater abstraction [5]. Other suitable areas include smaller community districts on the city edge outside the densely built up city areas, such as Dhulikhel, Sano Khokana, or Tokha.

2 WORK METHODOLOGY

Constructed wetlands is a biological wastewater treatment technology designed to mimic processes found in natural wetland ecosystems. These systems use wetland plants, soils and their associated micro-organisms to remove contaminants from wastewater [6].

Categorized by flow pattern, we divide constructed wetlands into free water surface wetlands and subsurface flow wetlands. Both these types of constructed wetlands utilize emergent aquatic vegetation. In the free water surface wetlands, water flows at a shallow depth horizontally over media which support the roots of the vegetation. The flow is above the substrate which makes the top layer of flow aerobic and the lower layers can be anaerobic. In addition to treating wastewater, free water surface wetlands can be aesthetically pleasing as they look like natural marshes and may provide wildlife habitat.

Subsurface flow wetlands are made up of pretty much the same components as the free water surface wetlands. The difference is that the flow of wastewater is designed to remain below the top of the media in the subsurface flow wetlands. Because the flow of wastewater is in the subsurface, odour and pest problems are minimized. Wastewater has also more contact with the porous media because the flow runs through the media and not over it. This can make the subsurface flow wetlands smaller in size compared to a free water surface wetlands treating the same amount of wastewater that is important for city areas.

Free water surface wetlands can become potential mosquito breeding grounds if not properly designed. Subsurface flow wetlands are therefore preferred in tropical and subtropical climates and are as well the best solution for Nepal, Kathmandu Valley, where the area available for their construction is rather limited.

There are two main directions of flow in the subsurface flow wetlands; horizontal flow and vertical flow, both has certain advantages and limitations. By combining them, we get a hybrid system that complements each other [7]. In horizontal flow wetlands, the wastewater flows from the inlet in the bed to the outlet of the bed in a horizontal path. As the wastewater moves slowly through the porous substrate, it encounters anaerobic, aerobic and anoxic zones. Organic pollutants are effectively removed in horizontal flow wetlands by microbiological degradation and by physical and chemical processes. The removal of nutrients is limited there due to lack of oxygen in the bed. Nitrates are however removed.

In vertical flow wetlands, the wastewater is fed intermittently over the bed so that it floods the top of the bed. The wastewater then moves vertically through the substrate before it is collected in drainage pipes at the bottom. Between loadings the bed is drained free of wastewater which allows air to fill the bed again. The next loading traps the air inside the bed and leads to good transfer of oxygen which allows nitrification.

Vertical flow wetlands removes organic matter and pathogens efficiently and takes up less space than a horizontal flow wetlands. Horizontal flow wetlands are better when it comes to removal of solids and as well they remove organic matter efficiently, but it takes up more space. Vertical flow wetlands have better oxygen transfer, hence the ability to nitrify, but can become clogged if the selection of media is wrong.

For Nepal, the horizontal flow constructed wetlands are easier to be maintained successfully, however the hybrid system could be better in the developed areas in the cities, with the limited space; for example for the private houses or schools.

Even though constructed wetlands have proven effective for treating different kinds of wastewater, there are still some challenges in the promotion of them. Despite constructed wetlands being a low-cost technology it might be difficult to convince people to take a basic care about the constructed wetlands and pay to treat their wastewater rather than just discharging it into the river (as there is no legislation and penalties to "convince" them). Gravel, sand and reeds might not be locally available for construction. Next, the management of the constructed wetlands is not organised by some communities in a satisfactory way, even the constructed wetland does not require much operation and maintenance. However it is important, that the system is treated efficiently and the constructed wetland is supervised by the responsible well-informed person from time to time.

In the literature and on available websites, about 60 constructed wetlands in the Kathmandu Valley and its immediate surroundings are marked. In the current situation of poor wastewater management, these wetlands are considered to be the most effective way of household wastewater treatment in this area.

During a field investigation directly in the Kathmandu Valley in September 2019, we were able to confirm only 23 of the documented constructed wetlands, the majority of which were described on the Environment and Public Health Organization (ENPHO) website (http://demo.crossovernepal.com/NP00100/?iec=factsheets).

In the field survey, we found that out of these 23 constructed wetlands, as few as seven constructed wetlands were fully functional and in good state. Three out of 23 constructed wetlands work only partially and 10 constructed wetlands have ceased to exist or have been destroyed by floods or landslides. In the case of three constructed wetlands, personal access to determine their situation was denied to us.

Considering the entities operating these 23 constructed wetlands, they can be divided in the following manner:

- Constructed wetlands on school premises: 1 functional, 1 partially functional, 2 not functional, 1 not accessible
- Constructed wetlands for bigger communities (over 20 families): 1 functional, 1 partially functional, 2 not functional, 1 not accessible
- Constructed wetlands for institutions:
  - ✓ Research institutions and offices: 2 functional, 4 not functional
  - ✓ Hospitals: 1 functional, 1 not functional (Figs 1 and 2: Sushma Koirala hospital constructed wetlands – the success story)
  - ✓ Monasteries: 1 not functional
  - ✓ Industrial establishment: 1 functional
- Constructed wetlands for private houses: 1 functional (other 2 functional, by hearsay), 1 partially functional, 1 not accessible.

The field research, that followed the literature review of the construction wetlands situation in the Kathmandu Valley, was organised via visits of the constructed wetlands and discussion with the owners/maintenance staff / community in the form of semi-standardised interviews (see the Fig. 2).

We focused especially on the socio-economic context of the constructed wetland implementation in Nepal, which, in absence of an effective state regulatory framework, is the most important factor for a successful implementation in situ. The questions of the interviews were related to:

Figure 1: Sushma Koirala hospital – the well-functioning, constructed wetland.

Figure 2: Interview with the person taking care of the Sushma Koirala constructed wetland.

- Ownership of the constructed wetland (who "owns it" – for whom it was constructed, why and how the particular community/settlement/household was identified for the constructed wetlands).
- Socio economic description of the community/settlement/household (in terms of caste, ethnic group, economic class), length of stay at the spot (permanent – immigrants), number of members/users.
- Involved stakeholders and financial source: who initiated wetland construction and why, what financial source were used and why these particular ones, etc.
- Why this particular technology of wastewater treatment was implemented? Was it promoted or supported by some financial schemes (national or international) or legislation?
- Present situation, technical management aspects, typology of the constructed wetlands, and the reasons for functioning, semi-functioning or not functioning of the constructed wetlands.

The results of our questionnaire campaign showed that the public awareness regarding constructed wetlands technology is the major challenge for this technology development. Based on our in situ investigation it is clear, that it is often difficult to involve community, institutions, and organizations for the installation of technology. The consideration of system as a low maintenance technology led to carelessness during operation and maintenance [8]. Further, wastewater treatment is not the priority for government, communities, and some institutions, due to the lack of strong legislation and standards [9]. (The detail evaluation of the interviews, however, will be the subject of the following paper, after the extensive campaign in the year 2020).

We were wondering why some constructed wetlands worked well and other had been neglected. The most important factor seemed to be the person/position of the operator, their awareness and level of involvement in the project. For this reason, we decided to place the pilot constructed wetland – which should be monitored from the beginning, to serve as a good practice example for potential technology transfer as well as training activities – either on school premises or on the premises of a big governmental or community institution.

For the implementation of the pilot constructed wetland, two potential sites were initially suggested – (a) NOBS schoolyard in Kathmandu, and (b) premises of the municipal office of Dhapakhel, one of the administrative units of the city of Lalitpur. The final decision was made when the Chairperson of Ward of Dhapakhel showed great interest in the collaboration and the project was also supported by the headmaster of the New Sumnima English School in Chakupat. This combination proved ideal for us.

3 PILOT PLANT

Unlike European households, Nepalese households produce little sewage. For this reason, horizontal system with subsurface flow was chosen for the project BIORESET model constructed wetland in Lalitpur. For the location of the constructed wetland, municipal office (Ward Office) of the Lalitpur Sub-metropolitan City, which is a state administrative authority, was chosen. This municipal office in Dhaphakhel suffers from (both sanitary and drinking) water supply shortage. Thus, the Chairperson of Ward welcomed the possibility of reusing the treated wastewater from toilets, washrooms and the kitchen (Fig. 3). At the same time, he suggested combining the wastewater treatment plant with a rainwater harvesting system, to assure enough water to maintain the wetland, which as he promised, was financed from the municipal budget. The Chairperson of Ward is a very active man. Among other things, he organises public health and environment protection promotion campaigns for the representatives of other municipal offices in Lalitpur. Thus, he will use the constructed wetland to raise awareness among his colleagues and, moreover, as a "learning aid" for the teachers and students of the New Sumnima School.

We placed the constructed wetland pilot plant on the premises of the municipal office and projected that a neighbouring police station is also connected to it; thus, the constructed wetland treats wastewater from the facilities of the municipal office, police station as well as a community training centre and it is designed for 22 to 25 users. The construction itself was carried out in September 2019 using funds provided by the company VODNÍ ZDROJE, a.s., the organisation AMBIS, the Ministry of Education, Youth and Sports of the Czech Republic, and the municipal office of Dhapakel.

The constructed wetland was located in the lowest part of the municipal office yard directly in front of the training centre building based on the agreement with the Chairperson of Ward, his team and the representatives of a local construction company.

The basic principle of the implemented system is horizontal flow of wastewater through a permeable substrate planted with wetland plants – calamus vegetation (Fig. 4). The sub-tropical climatic condition of urban areas of Nepal stimulated better growth of rhizosphere plant which led to better biological activity in soil for constructed wetlands performance [10], however in Dhapakel, we used recommended local plants (calamus).

The wetland is preceded by a septic tank where mechanical pre-treatment of sludge takes place. The filter bed of the wetland is 8 m long, 2 m wide and 1.5 m deep (Fig. 5), and is reinforced with a liner that was made by mixing a local material, i.e., clayey soil, with cement. This lining was thoroughly compacted both on the sides and the bottom of the filter bed. Then, filter material – gravel and, subsequently, sand – was placed into the filter bed. The treated water from the wetland is conveyed by gravity to a well 3 m deep, located at a distance of approx. 2 m (Fig. 6). This newly constructed well is used both for storage of sanitary water–treated water as well as rainwater, and has an infiltration purpose: through the bottom, the water infiltrates into the shallow aquifer, recharging the local drinking water sources.

The whole plant consisting of a constructed wetland, rainwater harvesting system and storage well was officially handed over for use to the Dhapakhel municipality in the presence of the headmaster of the New Sumnima School and the representatives of organisations involved in the project E! 12219 BIORESET: The use of controlled bioremediation for removal of specific types of contaminants.

The municipality Chairperson of Ward took over the constructed wetland appointing a person responsible for its maintenance and continuous inspection of the filter bed conditions, the biological components as well as the quality of water leaving the wetland.

Figure 3:   The Chairperson of the Ward welcomed the possibility of constructed wetlands implementation.

Figure 4:   Finalisation of the constructed wetland's site.

Figure 5:   Building of the constructed wetlands in Dhaphakhel, Lalitpur.

Figure 6: Building of the accumulation well.

Figure 7: Dhulikhel hospital – building of the new constructed wetlands.

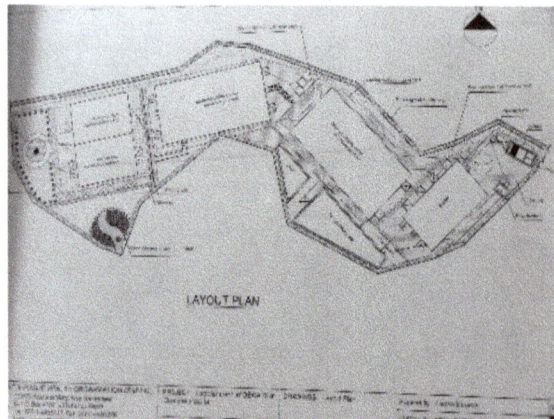

Figure 8: The plan for the new constructed wetlands at Dhulikhel hospital.

4 THE SPECIAL CASE: HOSPITALS

The implementation of constructed wetlands in private hospitals in the Kathmandu Valley is relatively successful. We would like to mention two success stories – Dulikhel hospital and Sushma Koirala Memorial Plastic and Reconstructive Surgery Hospital best practices.

In suitable hydrogeological and geographical conditions, private hospitals, in most cases also supported by foreign subsidies, are suitable for the implementation of constructed wetlands, as they are monitored both by the state and the donors as well as local communities as far as the quality of the released water and management of toxic substances are concerned; second, they have enough space for the installation of constructed wetlands and personnel to systematically deal with the operation of the constructed wetlands in the long run. For instance, in Sushma Koirala Memorial Plastic and Reconstructive Surgery Hospital located in Sankhu (where constructed wetlands were implemented in 2002 and enlarged and cleaned over time), the operation and maintenance of the plant is managed by the well – trained permanent hospital engineering staff and routine maintenance is included within the overall hospital maintenance plan.

The Dulikhel hospital (a Kathmandu University Teaching Hospital located in Dhulikhel Municipality, Kavrepalanchok District Nepal) had built the constructed wetlands in 1997 as a medium sized system consisting of three phase treatment (Anaerobic Baffle Reactor, Horizontal Wetland, Vertical Wetland) with two systems operating in parallel and a sludge drying bed to complete the wastewater treatment process. It had been renovated in 2008 (targeting 250 patients and staff of the hospital). The new reconstruction started in 2019 after serious damages by floods in the June 2019, because for the hospital, it is challenging to manage wastewater coming from the hospital and eliminate the organic pollutants from the water, which is used for the irrigation of agricultural fields in the vicinity (Figs 7 and 8).

However, until the time, nobody has raised the issue of inorganic (chemical and pharmaceuticals) pollution in the water coming from the hospital; there is no regular monitoring of these substances. The community people are happy to use water from outlet for irrigation purpose without the monitoring of the chemical parameters of the outlet water though twice in a year biological parameters are usually monitored.

The Dhulikhel hospital (that is the first subject who organised constructed wetlands in Nepal [11]. is promoting the nature-based technology approach because of following reasons: (1) they have enough land for constructed wetlands building (what is not a case in the more urbanised area of Kathmandu Valley), (2) there is suitable topography which doesn't require additional energy to run the system of constructed wetlands, (3) to promote locally available resources, (4) to show others the constructed wetland and their maintenance as a model, and (5) as this solution is economic in long term though it was comparatively quite expensive at the beginning (but there was the financial support by the international donor managed by ENPHO for the start-up activities).

Similar constructed wetlands can be replicated in other areas, not only for the hospitals, but as well for communities, thanks this "success story," as it has good network and information flow because of (a) many patients and their families/visitors can see and learn, (b) students are passing the training in the hospitals and could bring the constructed wetlands experiences to their home areas, (c) the Rural health camp is organised in the area and both, the Public health department and the Community development department can use the constructed wetlands case as teaching tool/subject and finally (d) many organised visitors (around 20 groups in a year) visit and see the system.

The horizontal flow constructed wetlands have long been used primarily for treatment of municipal or domestic wastewaters. Nowadays, they focus not only on common pollutants but also on special parameters available in the waters from agriculture or hospitals (such as pharmaceuticals, endocrine disruptive chemicals or linear alkylbenzensulfonates) [1].

However, the question remains whether the use of constructed wetlands for treating wastewater from hospitals in Third countries (that means as well as Nepal) is safe, because the management of toxic materials and residues is not satisfactorily dealt with within the wastewater management schemes. The quality of the water flowing into and released from the constructed wetlands is not regularly monitored for all the necessary parameters (such as various pharmaceuticals that can get into the water with the patients' urine).

Therefore, it is very likely that the water flowing out of the constructed wetlands may contain some Emerging Contaminants (synthetic organic chemicals that are detected in the natural environments), which are not degradable in the natural rock formations and can contaminate surface or ground water, or water used for irrigation of farmland.

For these reasons, we decided to establish a model constructed wetland pilot plant in the Kathmandu Valley in other environment than that of a hospital.

5 CONCLUSION

Due to the failure of the large treatment plants, small and decentralized treatment systems such as constructed wetlands are of high importance in Kathmandu Valley till now (the year 2019).

As discussed above, the most suitable environment for the constructed wetland pilot implementation has been identified in areas which have funds and staff for their operation, enough space as well as enough wastewater polluted with organic contaminants, and where the toxic or industrial pollutants are not expected to be presented in the wastewater.

At the same time, the operator of the constructed wetland (being also the manager of the premises where it is located) is interested in the environment and the abidance by the environment protection rules supported by his religious and social convictions as well as economic interests is a matter of prestige for him.

In private schools, for example, the presence of a constructed wetland may enhance the school's reputation, which will reflect in the students' increased interest in studying there and, at the same time, will ensure appropriate education of the students in the area of preserving a high-quality environment for a life (as an example, a very successful constructed wetland and rainwater harvesting system in the Satya Sai Sikshya Sadan school can be mentioned – see the Fig. 9). In another highly suitable type of site – municipal premises (such as Ward Offices) where the necessary funding for the operation of the constructed wetland can be procured more easily – the constructed wetland practice can be spread among other Chairpersons as well as communities living in the immediate vicinity thanks to a manager that is open to innovations.

Such communities, led by an enlightened leader who is in contact with "his" Chairperson of Ward, can then be effectively trained as to how to properly handle constructed wetlands, manage them and ensure their operation; via the municipality and its connection to appropriate national or international organisations, they can even obtain the necessary "initiating" means to launch the application of a constructed wetland in their own locality. Another suitable type of site, which is specific for Nepal and where a significant development in alternative tourism is now taking place, includes a constructed wetland implementation in tourist resorts or centres for spiritual practice and yoga for foreign clients, where it is desirable to demonstrate a connection to nature (as an example of partial success, the Namo Budha Resort can be mentioned – see Fig. 10).

Figure 9: Constructed wetland in the Satya Sai Sikshya Sadan school.

Figure 10: Semi functioning constructed wetland in the Namo Budha Resort.

Under the BIORESET project, we finally decided to establish a model constructed wetland on municipal land (Ward Office of Lalitpur Sub-metropolitan City), where sufficient amounts of wastewater as well as suitable space and a responsible person to oversee the operation of the constructed wetlands was ensured [12]. Strategical issue was as well to involve not only Ward Office of local government and police station to the activities, but as well the headmaster of the New Sumnima English School in Chakupat, living in the area and involved in the Ward Management Committee of the Municipality. The selection of the governmental place for the constructed wetlands building and operation and the educative and training potential have the long term positive impacts on the knowledge transfer and replication of the constructed wetlands in other areas/community places.

ACKNOWLEDGEMENTS
This paper was developed thanks to the EUREKA project E! 12219 BIORESET, co-financed by the Ministry of Education, Young and Sports, Czech Republic (50%) and organizations VODNÍ ZDROJE a.s. and AMBIS, a.s., as well thanks to the INTER-COST project "Natural Based Solutions for water management in cities", financed by the Ministry of Education, Young and Sports, Czech Republic.

REFERENCES

[1]   Vymazal, J., The use constructed wetlands with horizontal sub-surface flow for various types of wastewater. *Ecological Engineering*, **35**, pp. 1–17, 2009.

[2]   Green, H., Poh, S.-Ch. & Richards, A., Wastewater treatment in Kathmandu, Nepal. *Massachusetts Institute of Technology*, 2003. http://web.mit.edu/watsan/Docs/Student%20Reports/Nepal/NepalGroupReport2003-Wastewater.pdf.

[3]   Mishra, B.K. et al., Assessment of Bagmati river pollution in Kathmandu Valley: Scenario-based modelling and analysis for sustainable urban development. *Sustainability Water Quality and Ecology*, **9–10**, pp. 67–77, 2017. http://dx.doi.org/10.1016/j.swaqe.2017.06.001.

[4]   NEP, Kathmandu Valley Wastewater Management Project (2018): Kathmandu Upatyaka Khanepani Limited Project Implementation Directorate Anamnagar, Kathmandu. http://www.kuklpid.org.np/kukl/.

[5]   Boukalova, Z., Těšitel, J., Hrkal, Z. & Kahuda, D., Artificial infiltration as integrated water resources management tool. *Water Pollution XII, Water Pollution Conference*, The Algarve, Portugal, pp. 201–210, 2014.

[6]   Tuladhar, B., Shrestha, P. & Shrestha, R., *In Urban Sanitation, Decentralised Wastewater Management Using Constructed Wetlands*, ENPHO, pp. 86–94, 2008.

[7]   Yalcuk, A. & Ugurlu, A., Comparison of horizontal and vertical constructed wetland systems for landfill leachate treatment. *Bioresource Technology*, **100**, pp. 2521–2526, 2009. www.elsevier.com/locate/biortech.

[8]   Shrestha, R.R., Application of constructed wetlands for wastewater treatment in Nepal. Dissertation, University of Agricultural Sciences, 1999.

[9]   Pudasaini, K., Performance of wastewater treatment plants (BASP and SWTP) in Kathmandu valley: Case study of Bagmati area sewerage treatment plant (BASP) and Sunga wastewater treatment plant (SWTP), Master's thesis, Institute of Water Education, 2008.

[10]   Murthy,V. K., Khanal, S.N., Majumder, A.K., Weiss, A., Shrestha, D. & Maharjan, S., Assessment of performance characteristics of some constructed wetlands in Nepal, Kalmar Eco-Tech '07 Kalmar Sweden, 26–28 Nov., 2007.

[11]   Environment and Public Health Organization. www.enpho.org.

[12]   Boukalova, Z., Těšitel, J. & Gurung, D.B., Constructed wetlands and their implementation on private and public land in Kathmandu valley, Nepal. *WIT Transactions on Ecology and the Environment*, vol. 229, WIT Press: Southampton and Boston, pp. 1–8, 2019.

MEMBRANE BIOREACTOR WASTEWATER TREATMENT STRATEGIES FOR SENSITIVE COASTAL ENVIRONMENTS

MICHAEL C. GALLANT & MAURICE W. GALLARDA
Pluris Holdings LLC, USA

ABSTRACT
Southeastern North Carolina is predicted to experience an 80% population increase in the next 20 years, presenting challenges to water and wastewater infrastructure. Due to the topography and geology of the coastal plain environment, wastewater disposal can be problematic, historically utilizing large tracts of land for spray application or discharges directly into nutrient-sensitive waters. Previous treatment scenarios ranged from facultative lagoons to extended aeration processes. In order to meet the upcoming challenges posed by considerable growth, Pluris Holdings has embarked on designing, permitting and constructing a series of membrane bioreactor (MBR) wastewater treatment plants that produce effluents meeting groundwater and drinking water standards for nutrients. These plants have allowed for a wider array of disposal options requiring less land and capital costs than previous disposal systems, and they reduce the pollutant load to the receiving stream in the case of a direct discharge to coastal surface waters. Two methods for effluent disposal have been utilized in these plants. High rate infiltration (HRI) and national pollutant discharge elimination system direct discharge (NPDES). These methods reduce costs, require small amounts of land and responsibly deal with the disposal of wastewater. Due to the excellent quality of the MBR effluent, water reuse can also be utilized in conjunction with these two disposal methods. The use of MBR technology, innovative HRI disposal and/or NPDES disposal represent a paradigm change in waste-water treatment in the southeastern U.S. This change has required a close working relationship with the regulatory community. In this paper Pluris will discuss three separate projects including the design and permitting of each and the subsequent results, including effluent pollutant levels and reduced impacts on the environment.
Keywords: wastewater, MBR, membrane, nutrient, coastal North Carolina, NPDES, high rate infiltration.

1 INTRODUCTION
Southeastern North Carolina is expected to see a growth rate of 80% over the next 20 years. This environmentally sensitive area is the focus of this paper and the home to three advanced membrane bioreactor wastewater treatment facilities owned and operated by Pluris Holdings, LLC (Pluris), a privately held public utility.

Growth in southeastern North Carolina is driven primarily by a warm climate and rich environmental heritage. This rate of growth has been extremely challenging for previously rural municipalities that do not traditionally have the infrastructure or institutions capable of responding to the increased demand for services.

In the past, residents in the area relied heavily on primary treatment alternatives, such as individual on-site systems (septic tanks and nitrification fields), small package systems and larger primary treatments works consisting of facultative lagoons with spray application fields.

Historically, these systems were able to provide adequate treatment without undue stress on the local environment and surface waters. These surface waters have been routinely used for recreation, fishing and shellfish harvesting. They are a very large part of the coastal lifestyle and are prized by residents and visitors alike.

WIT Transactions on Ecology and the Environment, Vol 242, © 2020 WIT Press
www.witpress.com, ISSN 1743-3541 (on-line)
doi:10.2495/WP200121

The question presented to us is, how do local and regional governmental agencies absorb rapid growth and development without damaging the environment? Furthermore, how does a community create a sustainable model for wastewater treatment in such an environment?

In response to these challenges, Pluris has adopted membrane bioreactor (MBR) technology to achieve an excellent effluent quality that meets and/or exceeds groundwater and surface water standards for nutrients, primarily nitrogen and phosphorus.

Additionally, Pluris has pioneered the use of innovative in situ high rate infiltration (HRI) basins to dispose of effluent and create surplus groundwater that can be reused in the surrounding area without regulatory oversight.

2 DESCRIPTION OF THREE WASTEWATER TREATMENT FACILITIES

Pluris owns and operates three wastewater treatment plants in southeastern North Carolina. Fig. 1 shows the locations of the referenced membrane bioreactor wastewater treatment facilities in the coastal areas of southeast North Carolina.

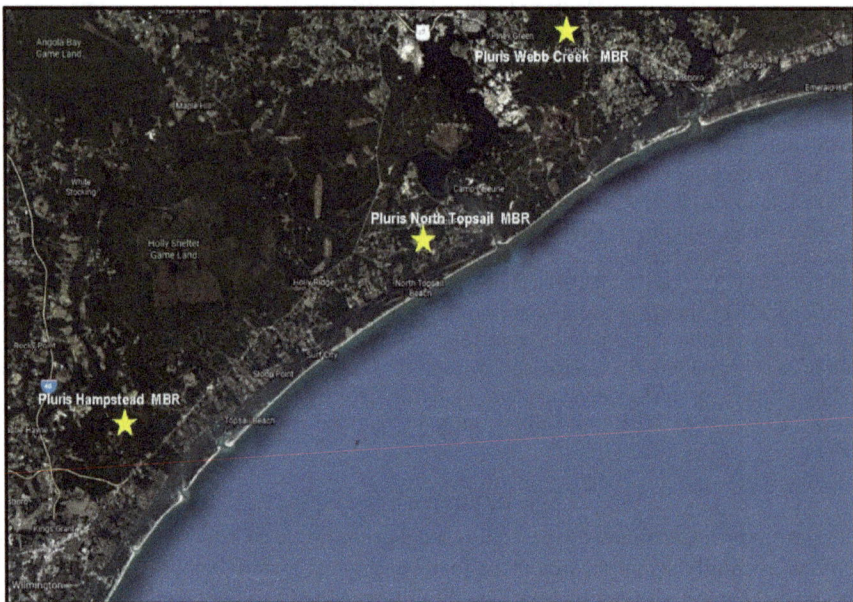

Figure 1: Locations of reference Pluris membrane bioreactor wastewater treatment facilities.

All of the systems are within 5 miles of the Atlantic Ocean, in North Carolina. These systems all border wetlands and/or tributaries of sensitive coastal waters. These waters range in state designations, from Class C to Class SA:HQW waters.

Class C waters are generally swamp waters and are not nutrient sensitive. Class SA:HQW waters are shell fishing waters and high quality waters used for recreational purposes. These waters can be nutrient sensitive, previously impaired and or primary nursery areas.

Historically, advances in treatment strategies have been focused on the regulatory landscape that existed at that time. Twenty-five years ago, most treatment plants were designed for 30/30 limits (30 mg/L BOD5 and 30 mg/L TSS). Today, treatment requirements have increased considerably. Pluris took the innovative step to provide treatment to levels below these requirements and to provide the highest treatment levels in the region. All of the facilities discussed in the paper utilize 0.04 micron membranes and are designed to achieve an effluent concentration of total nitrogen (TN) of 4 mg/L and a total phosphorus (TP) concentration of 2 mg/L. See Table 1 for the design, permit and typical standard limits for nitrogen and nitrate (see appendix and [1], [2]).

Table 1: Limits set by North Carolina Department of Environmental Quality (NCDEQ).

Parameter	Pluris North Topsail	Pluris Hampstead	Pluris Webb Creek
Design Limit, Nitrate (mg/L)	1.88	1.88	0.16*
Permit Limit, TN (mg/L)[1],[2],[3]	4	4	2*
Groundwater Standard Nitrate (mg/L)[4]	10	10	10
Surface Water Standard Nitrate (mg/L)[5]	10	10	10
* This limit is for Ammonia			

2.1 Facility 1: Pluris North Topsail wastewater treatment facility

The Pluris North Topsail system was purchased in 2008 from another private utility. At that time the plant provided treatment for 3,316,000 liters per day using three facultative lagoons in series. After the lagoon treatment the effluent was disinfected with a tablet chlorinator before being applied to 162 hectares of spray fields.

These spray fields bordered on Mill Creek which has a stream classification of SA:HWQ. The operation was land intensive, and the surrounding area was under a moratorium for new sewer connections. The permit limits did not protect the receiving stream from the subsequent runoff that periodically occurred from spray fields, especially during the wet periods of the year. The lagoons were subject to large impacts from tropical rain events such as hurricanes and tropical storms.

Pluris purchased the plant and embarked on building a 1 million gallon per day (MGD) MBR plant. This plant disposed of the effluent into two below ground level HRI basins with an innovative design that enhanced infiltration operation and maintenance of the basins. The key component of this design was an engineered material used in the center dyke construction of the basins. This material consists of a well-graded washed sand material with an angular shape that resists rilling and sloughing compared to in-situ soils or typical sands found in the area.

The new MBR plant and the two basins were constructed on less than 5 acres of land and are capable of treating 1 MGD of wastewater to tertiary standards. Compare this with the previous plant, which provided primary treatment and used over 400 acres of land.

This improvement in treatment had a significant impact on the area. First, it greatly reduced the mass of pollutants being applied to the environment. Second, it allowed for the moratorium on sewer connections to be lifted. When the plant was first purchased it served 2,000 customers. Today the customer base is over 6,000.

The values in Table 2 for the MBR mass loading were calculated using the maximum concentration per the permitted limit and the design flow. The mass loading for the lagoons was calculated using a typical result from a sample taken on 9 October 2019.

Another innovation designed into this plant was the use of a surplus groundwater impoundment. The HRI basins are surrounded by an underdrain with gravity drains to an 8-acre pond. This pond can be used as a source of groundwater for reuse in the community. Pluris has since constructed a lift station and reuse force main to distribute this resource to

Table 2: Pluris North Topsail yearly mass pollutant loading MBR versus lagoon treatment.

Treatment	Flow	TN load per year
MBR	1.0 MGD (3.785 MLD)	12,176 lbs (5,523 kg)
Faculative lagoons	0.876 MGD (3.316 MLD)	34,666 lbs (15,724 kg)

Figure 2: Amenity pond for aesthetics and water reuse.

the community. Currently, this system distributes groundwater to an amenity pond at a nearby subdivision (Fig. 2).

As this community is developed, this pond will become a centerpiece of the community, providing wildlife habitat as well as a pleasing and aesthetic water feature.

2.2 Facility 2: Pluris Hampstead wastewater treatment facility

The Pluris Hampstead wastewater treatment plant was constructed in response to calls from the development community for sustainable wastewater treatment. The Hampstead community has experienced double digit growth year over year. In 1993, the area had a population of 32,959. Today, the population is 54,820 [3].

This is an increase of 66% in 25 years. The projected population by 2038 is 86,078 [3]. This is an increase of 57% from the current population. Growth trends show that the majority of this growth will occur in the greater Hampstead area due to its proximity to the coast and the larger metropolitan area of Wilmington.

The previous treatment strategies were largely dependent on individual on-site systems. This limited development and has had a history of poor compliance and oversight. Furthermore, water tables in the area are very shallow with most areas having a seasonal high water table of less than 36" below the existing grade. Rising sea levels will likely increase the failure of these systems, especially those that are at lower elevations and nearer the coast.

Pluris Hampstead has provided a regional and sustainable solution to these problems. Like the other Pluris systems, the Hampstead facility utilizes 0.04 micron membranes. Unlike the Pluris North Topsail MBR plant, Pluris Hampstead was permitted without the use of any means of post-membrane disinfection. Citing the United States Environmental Protection Agency (EPA) white paper on the subject, the Pluris design team was able to show that MBR technology alone is sufficient for bacteriological disinfection [4], [5].

Pluris Hampstead is a conjunctive disposal plant that has both non-discharge and Federal EPA National Pollution Discharge Elimination System (NPDES) permits. The non-discharge permit specifies the use of two HRI basins using the innovative design used in the Pluris NT plant. The NPDES permit was added at the request of the regulatory agency and was permitted with disinfection only because an NPDES permit required it as a redundant system.

The Pluris Hampstead MBR plant is equipped with a peracetic acid disinfection system to meet the NPDES requirements. Pluris was the first wastewater treatment plant in the state to use this method of disinfection. Peracetic acid is a very strong disinfectant that does not cause the formation of disinfection by-products like chlorine. Capital costs are inexpensive compared to ozone or ultraviolet light systems, and the metering system uses far less electricity.

It is important to note that the NPDES discharge has never been used at the Pluris Hampstead MBR plant.

Like the Pluris North Topsail plant, the Hampstead MBR plant has a surplus groundwater impoundment and is owned by a third party, allowing the owner to resell the surplus groundwater to the surrounding planned subdivisions for irrigation.

It is important to note that the discharge parameters for the NPDES permit are far less restrictive that those for the non-discharge permit. At the Hampstead facility the receiving stream is classified as C Swamp waters. Table 3 lists the typical limits for various parameters based on the two disposal methods (see appendix).

Table 3: Pluris Hampstead permit limits, non-discharge versus NPDES.

Parameter	Units	Non-discharge limit	NPDES limit
BOD 5	mg/L	10	5
Fecal coliform	#/100 mL	14	200
Total nitrogen	mg/L	4	No limit
Total phosphorus	mg/L	2	No limit
TSS	mg/L	15	30

Although the receiving stream at the Pluris Hampstead MBR plant is not as environmentally sensitive as other areas, the engineering design team believed it was important to treat the effluent to the highest standards. This avoids issues with the plant operations and also ensures compliance in the case of future effluent restrictions or a reclassification of the receiving stream.

2.3 Facility 3: Pluris Webb Creek wastewater treatment plant

Pluris Webb Creek is located in Hubert North Carolina. Prior to Pluris purchasing the aging original sequential batch reactor (SBR) facility, it had been in operation for over 25 years. Due to the poor operating record of the previous utility owners, the utility was seized by the North Carolina Utilities Commission (NCUC), and Pluris was appointed as the emergency operator. Pluris subsequently purchased the utility with the commitment to design and build a new MBR plant.

The SBR plant, under the auspices of the original owners had a storied history with the North Carolina Department of Environmental Quality (NCDEQ) having incurred over 400 violation notices. The plant regularly exceeded its bacteriological permit limits. Out of the three facilities Pluris Webb Creek has the greatest potential effect on the coastal environment as it has a surface discharge NPDES permit, and the receiving stream is classified as SA:HQW.

In order to alleviate the future adverse environmental impacts, Pluris designed and is constructing a new 0.35 MGD MBR plant on the site. Like all other Pluris plants, this plant will utilize 0.04 micron membrane technology.

As of the writing of this paper, this plant is still nearing 50% completion. A recent drone photograph was taken showing an aerial view of the new MBR facility at Webb Creek (Fig. 3).

When complete this plant will eliminate the issue of bacteriological permit violations. It will also greatly reduce the concentrations of nitrogen and phosphorus in the effluent. Table 4 lists the design versus permit limits (see appendix).

Figure 3: Aerial drone photograph of new Pluris Webb Creek MBR treatment facility.

Table 4: NCDEQ Pluris Webb Creek permit limits, design versus permit limits.

Parameter	Units	Design	NPDES permit limit
BOD 5	mg/L	5	12/6*
Fecal coliform	#/100 mL	14	200
Total nitrogen	mg/L	4	No limit
Total phosphorus	mg/L	2	No limit
TSS	mg/L	5	30
*Winter/Summer			

3 PERMITTING CONSIDERATIONS

NPDES surface discharges are generally discouraged by the regulatory community in southeastern North Carolina. Several environmental groups regularly challenge applications for discharge permits including the North Carolina Coastal Federation, the White Oak River Keeper, the Cape Fear River Keeper and the North Carolina Shellfish Growers Association. These local environmentalists are concerned about the health of the coastal waters and strive to conserve not only the natural resources but the livelihood of the commercial fishing industry.

The Pluris team takes environmental concerns very seriously and works hard to find solutions that are reasonable, prudent, and environmentally responsible. The effluent from the three facilities discussed in this paper routinely have test results, as shown in Table 5 below showing pollutant levels as not detectable.

Table 5: Typical environmental laboratory test results.

Test	Results
Ammonia nitrogen	< 0.2 mg/L
Residual suspended (TSS)	< 2.5 mg/L
Total phosphorus	0.22 mg/L
BOD	< 2 mg/L
Nitrate nitrogen (calculated)	
Nitrite nitrogen	< 0.02 mg/L
Nitrite+nitrite-nitrogen	0.47 mg/L
Nitrate nitrogen	0.47 mg/L
Total nitrogen (calculated)	
Total Kjeldahl (TKN)	< 0.5 mg/L
Total nitrogen (calculated)	< 0.5 mg/L
Fecal coliform	2 MPN/100 mL

The NCDEQ is responsible for ensuring that the utilities meet all requirements for effluent parameters. All designs are reviewed and approved through the central office in Raleigh, North Carolina.

Due to the fact that Pluris regularly designs innovation into their plants, such as the referenced HRI basins and peracetic acid systems described above, permitting can be difficult. In some cases, the design team needed to meet with permit review staff and produce documentation and engineering data to show that technologies like MBR can be used with satisfactory results. MBR is used worldwide but is still relatively new in the U.S.

Just as the NCDEQ ensures that a plant will meet its permit limits, the NCUC regulates the utility such that rate payers are impacted as little as possible. Basically, the NCDEQ wants to ensure effluent quality regardless of cost, and the NCUC wants to make treatment as inexpensive as possible regardless of environmental impacts.

It is for this reason that Pluris regularly meets with the NCUC to discuss and review design plans prior to construction to avoid any issues.

4 CONCLUSION

All signs point to increasing regulatory oversight and tighter effluent restrictions for wastewater plants in southeastern North Carolina. The increasing population demands more services and infrastructure but at the same time desires that the natural heritage of the area be maintained.

Pluris has designed three treatment systems that meet or exceed regulatory limits and protect the surrounding environment. In some cases, reuse of surplus groundwater is available, increasing the sustainability of the system. There may come a time when reuse is mandated by the regulatory community. At that time, effluent quality will determine how that water can be reused. By exceeding standards for groundwater and surface water, Pluris has provided the most responsible and viable business model. Future considerations will most likely include an array of reuse options as well as indirect potable reuse.

REFERENCES

[1]   North Carolina Department of Environmental Quality (NCDEQ), 02L Groundwater Standards Table 5-21, 2013. https://files.nc.gov/ncdeq/documents/files/02L%20 Groundwater%20Standards%20Table%205-21%202013_0.pdf. Accessed on: 14 Dec. 2019.

[2]   North Carolina Department of Environmental Quality, NC Surface Water Quality Standards Table. https://deq.nc.gov/documents/nc-stdstable-06102019. Accessed on: 13 Dec. 2019.

[3]   North Carolina Office of State Budget and Management, Raleigh, NC. www.osbm.nc.gov/demog/county-projections. Accessed on: 15 Dec. 2019.

[4]   Francy, D.S. et al., *Quantifying Viruses and Bacteria in Wastewater – Results, Interpretation Methods, and Quality Control*, United States Geological Service, 2011.

[5]   Francy, D.S. et al., Comparative effectiveness of membrane bioreactors, conventional secondary treatment, and chlorine and UV disinfection to remove microorganisms from municipal wastewaters. *Water Research*, **46**, pp. 4164–4178, 2012.

APPENDIX

Figure A1: Pluris Hampstead non-disharge permit attachment "A". Available on request from author.

PPI 002 – Membrane Bio-Reactor Treatment System Effluent

	EFFLUENT CHARACTERISTICS		EFFLUENT LIMITS				MONITORING REQUIREMENTS	
PCS Code	Parameter Description	Units of Measure	Monthly Average	Monthly Geometric Mean	Daily Minimum	Daily Maximum	Measurement Frequency	Sample Type
00310	BOD, 5-Day (20 °C)	mg/L	10				3 x Week	Composite
00940	Chloride (as Cl)	mg/L					3 x Year[1]	Composite
31616	Coliform, Fecal MF, M-FC Broth, 44.5 °C	#/100 mL		14			3 x Week	Grab
50050	Flow, in Conduit or thru Treatment Plant	GPD	1,000,000				Continuous	Recording
00610	Nitrogen, Ammonia Total (as N)	mg/L	4				3 x Week	Composite
00625	Nitrogen, Kjeldahl, Total (as N)	mg/L					3 x Week	Composite
00620	Nitrogen, Nitrate Total (as N)	mg/L	10				3 x Week	Composite
00600	Nitrogen, Total (as N)	mg/L	4				3 x Week	Composite
00400	pH	su			6	9	5 x Week	Grab
00665	Phosphorus, Total (as P)	mg/L	2				3 x Week	Composite
70300	Solids, Total Dissolved – 180 °C	mg/L					3 x Year[1]	Composite
00530	Solids, Total Suspended	mg/L	15				3 x Week	Composite

1.   3 x Year sampling shall be conducting in March, July and November.

Figure A2: Pluris Hampstead non-discharge permit attachment "A". Available on request from author.

ATTACHMENT A - LIMITATIONS AND MONITORING REQUIREMENTS

Permit Number: WQ0037287

Version: 1.0

PPI 001 - WWTP Effluent

EFFLUENT CHARACTERISTICS		Units of Measure	EFFLUENT LIMITS				MONITORING REQUIREMENTS	
PCS Code	Parameter Description		Monthly Average	Monthly Geometric Mean	Daily Minimum	Daily Maximum	Measurement Frequency	Sample Type
00310	BOO, 5-Day (20°C)	mg/L	10				2 x Month	Composite
00940	Chloride (as Cl)	mg/L					3 x Year¹	Composite
31616	Coliform, Fecal MF, M-FC Broth, 44.5°C	#/100mL		14			2 x Month	Grab
50050	Flow in Conduit or thru Treatment Plant	GPO	50,000				Continuous	Recorder
00610	Nitrogen, Ammonia Total (as N)	mg/L	4				2 x Month	Composite
00625	Nitrogen, Kjeldahl, Total (as N)	mg/L					2 x Month	Composite
00620	Nitrogen, Nitrate Total (as N)	mg/L	10				2 x Month	Composite
00600	Nitrogen, Total (as N)	mg/L	4				2 x Month	Composite
00400	pH	SU			6	9	5 x Week	Grab
00665	Phosphorus, Total (as P)	mg/L	2				2 x Month	Composite
70300	Solids, Total Dissolved - 180°C	mg/L					3 x Year¹	Composite
00530	Solids, Total Suspended	mg/L	15				2 x Month	Composite

1. 3 x Year sampling shall be conducted in March, July and November

Figure A3: Pluris Hampstead NPDES permit page 3 of 7, Available on request from author.

Permit NC0089524

A. (1) EFFLUENT LIMITATIONS AND MONITORING REQUIREMENTS
[I5 A NCAC 02B .0400 et seq., 02B .0500 et seq.]

a. During the period beginning on the effective date of the permit and after receipt of the signed Engineering Certificate indicating completion of construction, and lasting until expiration, the Permittee is authorized to discharge treated domestic wastewater from *Outfall 001*. Such discharges shall be limited and monitored [1] by the Permittee as specified below:

EFFLUENT CHARACTERISTIC	DISCHARGE LIMITATIONS		MONITORING REQUIREMENTS		
	Monthly Average	Daily Maximum	Measurement Frequency	Sample Type	Sample Location
Flow	0.25 MGD		Continuous	Recording	Effluent
BOD, 5-Day, 20°c	5.0 mg/L	7.5 mg/L	3/Week	Composite	Effluent
Total Suspended Solids	30 mg/L	45 mg/L	3/Week	Composite	Effluent
NH3-N (April I – October 31)	I.1 mg/L	5.5 mg/L	3/Week	Composite	Effluent
NH3-N (November I -March 31)	2.5 mg/L	12.5 mg/L	3/Week	Composite	Effluent
Fecal Coliform (geometric mean)	200/100 mL	400/100 mL	3/Week	Grab	Effluent
pH	Not less than 6.0 S.U. nor greater than 9.0 S.U.		3/Week	Grab	Effluent
Dissolved Oxygen	Not less than 5.0 mg/L, daily average		3/Week	Grab	Effluent
Dissolved Oxygen, mg/L [2]			Weekly	Grab	Upstream & Downstream
Total Residual Chlorine [3]		17 µg/L	Weekly	Grab	Effluent
Temperature, °C	Monitor and Report		Daily	Grab	Effluent
Temperature, °C [2]	Monitor and Report		Weekly	Grab	Upstream & Downstream
Total Nitrogen, mg/L [4]	Monitor and Report		Quarterly	Composite	Effluent
Total Phosphorus, mg/L	Monitor and Report		Quarterly	Composite	Effluent
Chronic Toxicity [5]			Quarterly	Composite	Effluent

Footnotes:

1. No later than 270 days from the effective date of this permit, begin submitting discharge monitoring reports electronically using NC DWR's eDMR application system. See special condition A. (4).
2. Upstream approximately 50 feet from the outfall and downstream approximately 870 ft from the outfall, at locations approved by Wilmington Regional Office. All instream samples shall be collected during a discharge event.
3. Total Residual Chlorine (TRC) limit and monitoring only apply if chlorine or chlorine derivative are used for cleaning the MBR units and is in contact with the wastewater. When required the facility shall monitor and report all effluent TRC values reported by a NC certified laboratory including field certified. However, effluent values below 50 µg/L will be treated as zero for compliance purposes.
4. Total Nitrogen (TN) == (N02-N + N03-N) + TKN, where (N02-N + N03-N) and TKN are Nitrite/Nitrate Nitrogen and Total Kjeldahl Nitrogen respectively.
5. Chronic Toxicity *(Ceriodaphnia)* @ 88.6%; February, May, August and November, see special condition A. (2).

b. There shall be no discharge of floating solids or visible foam in other than trace amounts.

Page 3 of 7

Figure A4: Pluris Webb Creek NPDES permit attachment "A". Available on request from author.

Permit NC0089877

A. (2.) EFFLUENT LIMITS AND MONITORING REQUIREMENTS
[15A NCAC 02B.0400 et seq., 02B.0500 et seq.]
Grade III Biological WPCS [15A NCAC 08G .0302]

During the period beginning upon the submittal of an Engineer's Certificate and lasting until expiration, the Permittee is authorized to discharge treated wastewater from outfall 001. Such discharges shall be limited and **monitored**[1] by the Permittee as specified below:

EFFLUENT CHARACTERISTICS	EFFLUENT LIMITS			MONITORING REQUIREMENTS		
Parameter Description – eDMR code	Monthly Average	Daily Maximum	Unit of Measure	Measurement Frequency	Sample Type	Sample Location[1]
Flow, in conduit or thru treatment plant [50050]	0.350		MGD	Continuous	Recorder	Influent or Effluent
BOD, 5-Day (20 Deg. C) [CO310] - Winter	10.0	15.0	mg/L	3 / week	Composite	Effluent
BOD, 5-Day (20 Deg. C) [CO310] - Summer	5.0	7.5	mg/L	3 / week	Composite	Effluent
Solids, Total Suspended [CO530]	30.0	45.0	mg/L	3 / week	Composite	Effluent
Nitrogen, Ammonia Total (as N) [CO610] - Winter	2.0	10.0	mg/L	3 / week	Composite	Effluent
Nitrogen, Ammonia Total (as N) [CO610] - Summer	1.0	5.0	mg/L	3 / week	Composite	Effluent
Temperature, Water Deg. Centigrade [00010]			deg. C	Daily (5/week)	Grab	Effluent
DO, Oxygen, Dissolved[2] [00300]			mg/L	3 / week	Grab	Effluent
Phosphorus, Total (as P) [CO665]			mg/L	Weekly	Composite	Effluent
Nitrogen, Total (as N) [CO600]			mg/L	Quarterly	Composite	Effluent
Fecal Coliform [31616]	200/100 mL	400/100 mL	cfu/100ml	3 / week	Grab	Effluent
pH [00400]	Between 6.8 and 8.5 Standard Units		s.u.	3 / week	Grab	Effluent
pH [00400]	Monitor and Report		s.u.	Weekly	Grab	Upstream & Downstream
Temperature, Water Deg. Centigrade [00010]	Monitor and Report		deg. C	Weekly	Grab	Upstream & Downstream
DO, Oxygen, Dissolved [00300]	Monitor and Report		mg/L	Weekly	Grab	Upstream & Downstream

*Winter: November 1 - March 31, *Summer: April 1 – October 31

Footnotes:
1. Upstream = at least 100 feet upstream from the outfall. Downstream = at the nearest road.
2. The minimum daily Dissolved Oxygen effluent concentration shall not be less than 6.0 mg/L.
4. Total Residual Chlorine (TRC) limit and monitoring only apply if chlorine or chlorine derivative are used for cleaning the MBR units and is in contact with the wastewater. When required the facility shall monitor and report all effluent TRC values reported by a NC certified laboratory including field certified. However, effluent values below 50 μg/L will be treated as zero for compliance purposes.

There shall be no discharge of floating solids or visible foam in other than trace amounts.

SECTION 5
SUSTAINABLE URBAN
DRAINAGE SYSTEMS

NINE STRATEGIES TO PROTECT AND ENHANCE WATERWAYS

GLENN BROWNING
Healthy Land and Water, Australia

ABSTRACT

In Queensland, the State Government legislation requires all new development to reduce stormwater pollutant loads discharging to the creek (in Brisbane for example, total suspended solids needs to be reduced by 80%, total phosphorus by 60% and total nitrogen by 45%). These simple targets have driven millions of dollars of investment throughout the State. Healthy Land and Water (HLW), a Brisbane-based natural resource management group were given the opportunity to review these stormwater design objectives for the State Government. Whilst it is acknowledged that they are a much-needed step to limiting our ecological impact, our research has raised questions about whether this is the best way of managing our waterway assets. By assessing the stormwater pollution targets using a first principles risk analysis, it was shown that there are fundamental inefficiencies with this approach. The current approach places no incentive on hazard minimisation or avoidance, it makes no distinction as to the downstream waterway value and it provides no opportunity to invest in value restoration, reconnection or conservation efforts. To address these issues HLW have created Strategic Waterways, a tool for categorising and prioritising waterway investments. The tool uses a risk/benefit model to assess, diagnose and then plan the treatment of various waterway ailments. Most importantly it allows for nine unique strategies to managing waterway value where previously there was only one or two. Strategies include: (1) value conservation, (2) value protection, (3) hazard minimisation, (4) risk mitigation, (5) value reconnection, (6) value creation, (7) gamechangers, (8) strategic offsets, and (9) minimum requirements. This paper explains the practical differences in these approaches and empowers waterway managers to build a balanced portfolio of waterway investments to create the biggest possible ecological return on investment.
Keywords: stormwater, waterways, management, strategy, pollution, risk, values, prioritisation.

1 INTRODUCTION

Currently the development industry in Queensland, Australia has one typical strategy for managing stormwater pollution; stormwater runoff from new housing estates is filtered via bioretention basins (vegetated sand filter beds) before it discharges to the drainage system and receiving waterways. Although this provision reduces potential environmental harm it does not consider the health condition of the local waterway [1] nor the extensive legacy issues across our urban waterways [2]. It also does not allow for hazard minimisation nor waterway improvement or value creation strategies [3], [4]. Monitoring of the waterways [5] suggests there is much work to do to ease current waterway pressures and we are further adding to their pressures through urban densification and climate change. Essentially, as an industry we are failing to achieve the objectives of the legislation under the Environmental Protection (Water) Policy [6] to "protect and enhance water values", After reviewing the science and the practice, HLW have found more can be achieved by taking a strategic approach to waterway management. This involves directing proportionate and appropriate effort and investment to locations where it can create the biggest impact. In circumstances where there is limited environmental funding, this approach to maximise ecological and social return on investment (ROI) is imperative.

1.1 Methodology for understanding the problem

HLW have previously assessed and documented the state of waterway management in Queensland in 2014 and again in 2017 [7]. More recently, HLW and Alluvium undertook an extensive review of scientific literature [1] throughout Queensland and internationally focusing particularly on the hydrologic and water quality impacts of urban development on waterways. Informed by this research, the team also held eight half-day workshops with industry leaders and key councils across the state [2] to gather valuable insight and a further appreciation into the current science and stormwater management practices across the State.

1.2 Proposed solution

To diversify the portfolio of activities funded and improve our ecological return on investment, HLW has created a new decision support tool to improve waterway condition called Strategic Waterways that has a 5-step process:

1. Condition assessment: *What is the current health status of the waterway?*
2. Diagnosis: *What is causing poor waterway health?*
3. Treatment strategy: *What treatment options should be used?*
4. Triage: *Which sections of the waterway has the highest priority?*
5. Monitoring: *How effective is our strategy?*

   This paper expands on and develops the theory behind each of the nine treatment strategies (i.e. Step 3) and how each strategy can be used most efficiently and effectively.

1.3 Target audience

This highly adaptable tool can potentially be used to assess many other types of engineering risks, such as safety, bushfire and heat risk, natural areas management, and flood management. HLW's Strategic Waterways tool will be useful for a variety of decision makers in local governments including:

- Waterway and catchment managers and planners;
- Stormwater drainage engineers;
- Asset managers, budgeting and finance.

1.4 How does the strategic waterways tool work?

Strategic Waterways provides users with a questionnaire about hazards "R", values "G" and needs "B". It then classifies a given waterway with a "RGB" colour code. A score of 0 indicates a low score and 250 indicates a high score. Since "RGB" colour coding can visually represent a diverse range of colours the system is quite versatile. Details of the colour coding system are outlined below:

- "Red" indicates how hazardous the catchment is;
- Where a hazard overlaps with a value it forms a risk ("Yellow");
- "Green" indicates how valued the catchment/waterway is;
- Where a value overlaps with a need it forms an opportunity ("Cyan");
- "Blue" indicates where there is the opportunity to recover or enhance value;
- Where a hazard can fulfil a need it forms a gamechanger ("Magenta").

The unique colour coding system means it is easy to visualise the combination of waterway hazards, values and needs. Once an "RGB" colour code is determined this helps diagnose what the waterway actually needs, and a corresponding treatment strategy can be assigned (Fig. 1). The scoring can also be used to triage and prioritise as well as keep track of improvement strategies. The Strategic Waterways [8] questionnaire and user guide can be downloaded for free from: www.waterbydesign.com.au/resources.

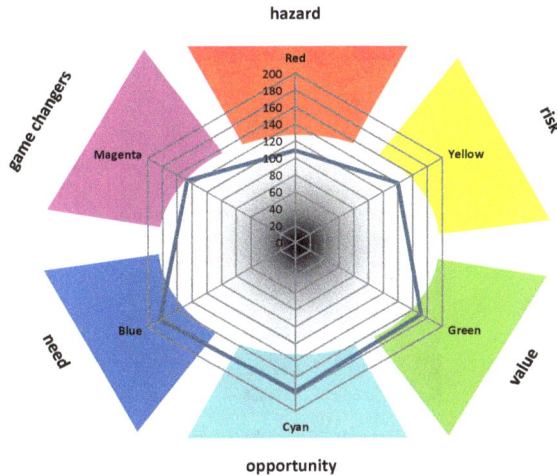

Figure 1: Hazard/value/need plot.

1.5 Treatment strategies

Once a waterway diagnosis has been undertaken, a waterway manager can start to assign an appropriate treatment strategy for each section of the waterway. For some pristine waterways, this strategy will focus on conservation, whereas for others the focus may be on hazard reduction. This remainder of this paper concentrates on the nine treatment strategies to improve waterway management including:

- Strategy 1: Value conservation;
- Strategy 2: Value protection;
- Strategy 3: Hazard minimisation;
- Strategy 4: Risk mitigation;
- Strategy 5: Value reconnection;
- Strategy 6: Value creation;
- Strategy 7: Gamechangers;
- Strategy 8: Strategic offsets;
- Strategy 9: Minimum requirements.

The nine strategies detailed above represent a marked improvement in the number of options available to stormwater managers under current Queensland stormwater quality regulations. Taking this nuanced management approach will account for the unique condition, pressures and prospects of each waterway and allow for better investment than standard approaches [8].

2 STRATEGY 1: VALUE CONSERVATION

Strategy aims:	To maintain existing values;
	To manage potential threats;
	To inspire further conservation action.
Qualifying criteria:	Waterways and catchments are in pristine condition.
Site profile:	"Eden" (Fig. 2).
Examples:	Daintree Rainforest, Noosa River.
"Green"	RGB Code: R = 0, G = 250, B = 0.

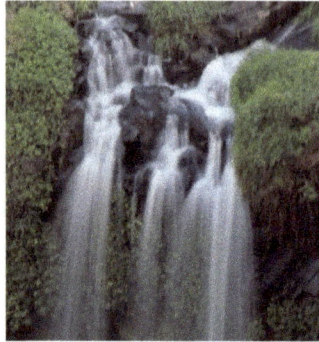

Figure 2: "Eden".

2.1 Strategy 1a: Manage generic threats

There is a need to underpin resilience of very high value waterways, that is we need to improve the ability to adapt or bounce back from shocks including:

- Climate impacts (e.g. temperature, flood, drought, bushfire);
- Invasive species (e.g. weeds and exotic fish).

2.2 Strategy 1b: Protect from future development

These waterway sites are to be mapped as protected areas under local government planning schemes and its catchments should not be cleared or developed.

2.3 Strategy 1c: Safeguard perceived value and justify protections

There is a risk that if these sites are locked up and hidden from view then they will be forgotten about and neglected. So, there is a need to continually demonstrate the value of pristine waterways to society and governments so that they continue to enact policies to keep them in their wild state. This can be achieved through scientific research, education and media. This will help safeguard against government policy changes and future catchment development.

2.4 Strategy 1d: Celebrate intrinsic value

By permitting opportunities for society to access and enjoy pristine waterway sites we can improve our understanding of the natural environment and our nature connectedness and this may inspire patrons to undertake conservation action. Collecting visitor numbers and visitor feedback will also help to justify continual investment in these wild areas.

3 STRATEGY 2: VALUE PROTECTION

Strategy aims:	To maintain existing waterway values;
	To manage potential threats;
	To permit a low-impact form of development.
Qualifying criteria:	Waterways have high value scores,
	Small portions of the catchment may already be cleared of vegetation and have development potential.
Site profile:	"The Shire" (Fig. 3).
Examples:	Gold Coast Hinterland (e.g. Currumbin Ecovillage).
"Green"	RGB Code: R = 0, G = 200, B = 0.

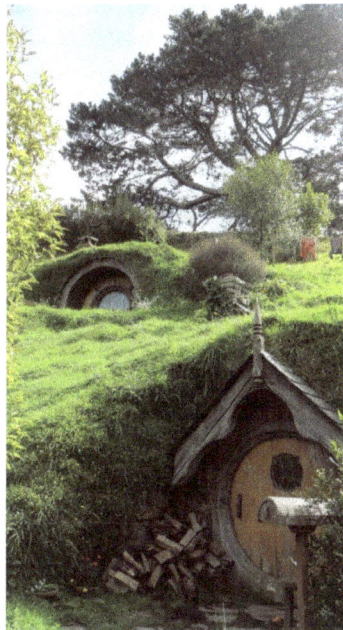

Figure 3: Green roofs as represented in the Hobbit "Shire". *(Photo: Sheila Thomson.)*

3.1 Strategy 2a: Manage generic threats

There is a need to underpin resilience of very high value waterways, that is we need to improve the ability to adapt or bounce back from shocks including:

- Climate impacts (e.g. temperature, flood, drought, bushfire);
- Invasive species (e.g. weeds and exotic fish).

3.2 Strategy 2b: Insulate waterways from development threats

Where waterways have high value, but their catchments can still be developed, waterways should be protected at the planning stage with very strict controls on any development.

Development within these catchments should have a "neutral or beneficial impact" on the local waterway and should employ low-impact design. The ideal development would be almost completely vegetated and would represent best practice in stormwater management with features such as rainwater tanks, permeable driveways and roads and green roofs.

4 STRATEGY 3: HAZARD MINIMISATION

Strategy aims: To control hazards at their source.
Qualifying criteria: Catchments with high pollutant loads.
Site profile: "A catchment detox".
Examples: Industrial areas such as Pinkenba, Port of Brisbane, Rocklea, Wacol, Archerfield, Slacks Creek and Underwood.
"Red"   RGB Code: R = 250, G = 0, B = 0.

To devise an adequate pollution abatement program, we need to understand each of the relevant pollutants as described below.

Table 1: Understanding waterway hazards.

Take an inventory of hazards	Origins: industrial, commercial, residential or agricultural Distribution/prevalence: via GIS Mapping
Understanding the hazard	Toxicity/harm Magnitude and recurrence of pollution events: acute/ chronic
Assessment of threat level	Priority should be given to the highest combined toxicity and prevalence
Management of threat	Can the hazard be practically eliminated or is containment the best that can hoped for?

4.1 Strategy 3a: Reduce pollutant toxicity

The toxicity of pollutants can be reduced via pollutant avoidance, substitution or control.

Legal controls are quite effective for industrial discharge and commercial products. Acute hazards such as DDT may need a direct ban. These harmful products can often be substituted with less harmful chemicals such as waterway friendly herbicides and pesticides.

Planning controls are quite effective at reducing the toxicity of stormwater discharge for residential areas. Setting water quality targets ensures stormwater is filtered prior to discharge to the creek which reduces contaminant concentrations and toxicity.

4.2 Strategy 3b: Reduce pollutant prevalence

The prevalence of waterway hazards and pollutants can be reduced via behaviour change programs and design controls.

Where the hazard is widespread there will be a need to invest in *behaviour change programs*, e.g. reduction of single-use plastic drink bottles, bags and straws.

Proactive *environmental design* in residential areas can help to avoid the creation and prevalence of stormwater pollution. Options include passive irrigation, bioretention, rainwater tanks, infiltration systems, permeable driveways and green roofs.

5 STRATEGY 4: RISK MITIGATION

Strategy aims:	To reduce harm to waterways through hazard mitigation.
Qualifying criteria:	Sites where a hazard flows to an area of high value forming a critical vulnerability.
Site profile:	"The Flash Point" (Fig. 4).
Examples:	Bells Creek, Mango Hill.
"Yellow"	RGB Code: R = 250, G = 250, B = 0.

Figure 4: Pristine waterways become polluted with construction sediment.

GIS mapping can highlight key locations where there is an intersection of values and hazards i.e. hotspots. By understanding the risk pathway [9], we then can identify the best strategies and management actions to break the risk path and protect our waterway values (Fig. 5).

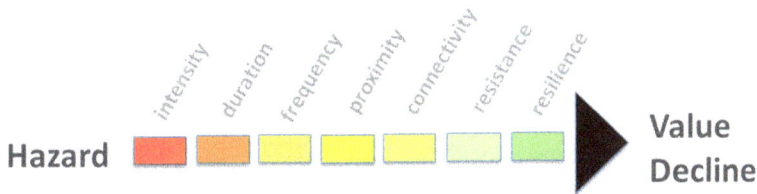

Figure 5: The risk pathway [9].

- Strategy 4a: Reduce hazard intensity, e.g. reduce pollutant toxicity (refer also Strategy 3).
- Strategy 4b: Reduce the duration of a hazardous event, e.g. provide sediment containment basins at construction sites.
- Strategy 4c: Reduce the frequency of a hazardous event, e.g. provide spill containment basins at on major roads.
- Strategy 4d: Reduce proximity of hazard to value, e.g. ensure adequate riparian buffer widths.

- Strategy 4e: Reduce connectivity of hazard to value, e.g. disconnect downpipes from the stormwater network.
- Strategy 4f: Improve the resistance of the value to a hazard, e.g. provide rock armouring to creek banks to prevent scouring.
- Strategy 4g: Improve the resilience of the value and allow opportunities for the waterway to self-restore, e.g. improve the flushing of a lake or pond (refer also Strategy 1).

6 STRATEGY 5: VALUE RECONNECTION

Strategy aims:	To expand this value spatially;
	To link to other value regions; or
	To improve complimentary value sets due to synergies.
Qualifying criteria:	Waterways must have a section with high value and opportunity to expand or improve value.
Site profile:	"The Missing Link".
Examples:	Davidson Street Project [10], Bancroft Weir, Barry's Weir.
"Cyan"	RGB Code: R = 0, G = 250, B = 250.

GIS mapping can help to locate areas with high recovery potential. The first step is to identify areas with high waterway value (refer also Strategy 1). Any adjacent areas need to be examined to identify opportunities to expand this value via the following sub strategies.

6.1 Strategy 5a: Value expansion

Where there are opportunities to expand outwards from existing nodes of high value, creation of new habitat areas will encourage local wildlife populations to expand into these new areas.

6.2 Strategy 5b: Value reconnection

Managing connectivity is identified as one of the key factors to improve resilience [11]. It has been shown that fragmentation of habitat can harm regional biodiversity. This strategy may be applied if there are opportunities to link several existing value nodes and create wildlife corridors. For example, there may be opportunity to construct a fishway or to reengage a floodplain that has previously been disconnected from a waterway.

6.3 Strategy 5c: Complimentary functions

Where there are potential complimentary or synergistic functions within existing high value nodes, there may be opportunity to improve the triple-bottom-line value of a waterway. For example, a new eco-tourism venture may be compatible with a national park or conservation area. For this to occur there needs to be an intersection of high value and a key need.

6.4 Strategy 5d: Value restoration

With a little seed funding it may be possible to harness nature's ability for self-restoration. A famous example of this is the reintroduction of apex predators to national parks which caused widespread improvements in the landscape. Investment into community-based behaviour change mechanisms can create chain reactions leading to greater environmental improvements.

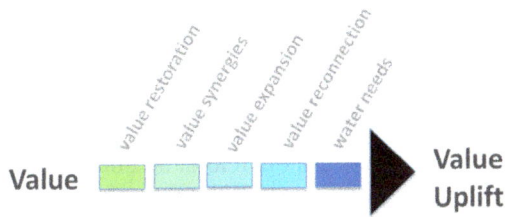

Figure 6: The opportunity pathway.

7 STRATEGY 6: VALUE CREATION

Strategy aims:	To meet the immediate needs of the community and the local ecology.
Qualifying criteria:	Catchments where there is little to harm in the way of existing waterway value.
Site profile:	"The Blank Canvas".
Examples:	Mackay's Cane Drains [12], Small Creek – Raceview [13].
"Blue"	RGB Code: R = 0, G = 0, B = 250.

Since these sites are usually isolated from other areas of value, there may be a lower chance that these projects catalyse further value increases. These projects should therefore be evaluated on their own merit and may have limited benefit to adjacent catchments.

7.1 Strategy 6a: Novel waterways

With a blank canvas, Local Governments can create the exact waterways they want to fulfil a community or ecological need. The city of Mackay is surrounded by cane fields. As the urban footprint expands there is opportunity to recreate "natural" waterways from the highly modified cane drains (see Fig. 7) [12]. Ipswich City Council has invested considerable sums to re-naturalise a section of Small Creek that had been cleared of vegetation, straightened and hardened via a concrete channel [13]. These creek naturalisation projects create new opportunities for residents to interact with nature. They also provide new habitat for local wildlife.

Figure 7: Cane Drain, Mackay.

7.2 Strategy 6b: Establishing refuges

We can improve the survival chances for migratory species (e.g. birds) by creating high quality refuges (e.g. wetlands) in areas where there may be little surrounding value.

It may also possible to create refuges out of man-made lakes or wetlands and undertaking fish restocking programs. These areas can often form an important insurance policy against degradation and deterioration of waterways in other areas of the State.

8 STRATEGY 7: GAMECHANGERS

Strategy aims:	To use a waterway hazard to fulfil a community or ecological need.
Qualifying criteria:	Sites need to produce a suitable form of waterway hazard; There needs to be a mechanism to convert this hazard into a useful product; There must also be a particular demand for the product close by.
Site profile:	"Turning Trash into Treasure".
Examples:	Redbank Plains Recreational Reserve [14], Jim Donald Park.
"Magenta"	RGB Code: R = 250, G = 0, B = 250.

8.1 Strategy 7a: Gamechangers

Gamechangers are usually technology-based solutions and are underpinned by the principles of the Circular Economy where waste products are converted into a resource. Essentially it mimics nature where waste products are decomposed and form the building blocks of the next lifeform for example fallen leaves decompose and become fertiliser for the next growth cycle. This strategy may not be feasible in all circumstances but if these types of solutions are found they can neatly address a waterway hazard and provide additional benefits to the community or local environment.

Urban development can increase the amount of water flowing to creeks leading to bed and bank scour. However, there is a real need to diversify water resources in a changing climate. Stormwater can help displace the need for drinking water and can be used to irrigate lawns and gardens. There are also opportunities to harvest recycled rainwater for flushing toilets and laundries. As illustrated (Fig. 8), there are several prerequisites that are required in order to achieve successful game-changing outcomes.

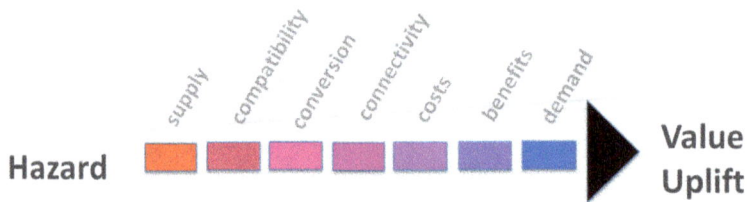

Figure 8: The gamechanger pathway.

The first prerequisite is to establish *compatibility*. Is there is a community, economic or environmental need for a particular hazard (e.g. can excess stormwater be used for irrigation)? The second prerequisite is to find locations where these hazards and needs intersect (*Connectivity*). *Supply* and *Demand* considerations need to be assessed. Stormwater harvesting systems cannot function at full productivity without a demand for non-potable water (e.g. an adjacent sports field to irrigate). The last prerequisite is to achieve an appropriate *Cost/Benefit* ratio (e.g. harvested stormwater is often too expensive when used as a drinking water supply due to high treatment and storage costs).

Other examples of gamechangers include drinking container deposit schemes (these fulfil economic needs as well as pollutant reduction) and using treated wastewater to recharge groundwater aquifers to prevent seawater intrusion into groundwater dependant ecosystems (these fulfil ecological needs as well as pollution reduction).

9 STRATEGY 8: STRATEGIC OFFSETS

Strategy aims:	To redirect finance to areas where they will have a much bigger impact.
Qualifying criteria:	The donor site should not discharge to a high value waterway; It should be relatively stable (i.e. established urban areas); Adjacent waterways should have limited prospects for future recovery.
Site profile:	"The Underperformer".
Examples:	Port of Brisbane [15], infill development in the Lower Norman Creek and Kedron Brook catchments.
"Grey"	RGB Code: R = 100, G = 100, B = 100.

There are four key scenarios where an offset can be legitimised.

9.1 Strategy 8a: Waterways unlikely to improve from investment

Where stormwater discharges directly to open ocean the benefits of a bioretention basin are unlikely to be as beneficial as in other circumstances. For example:

- Nutrient is managed by marine environment;
- Sediment is deposited on the sea floor;
- Flow change is unnoticed.

In this instance the same investment is best made upstream where it can have a bigger impact.

9.2 Strategy 8b: Higher than average infrastructure cost

There is a case for employing stormwater offsets in situations where a high percentage of money goes to ancillary works that are unrelated to stormwater treatment. For example, retaining structures may be required around bioretention basins on sites with steep land slope and services may require substantial relocating in retrofit applications at marginal ecological return on investment. If the same money was spent elsewhere it could have a much bigger impact.

9.3 Strategy 8c: High missed opportunity cost

Sometimes a proposed stormwater treatment measure will require allocation of too much premium land and another project on the same site would create a large economic return [15]. In this example, funds raised from commercial ventures in the CBD could contribute to a larger impact outside of the city CBD.

9.4 Strategy 8d: Insufficient funding for high impact projects

Where high impact projects (e.g. regional wetlands) have been identified in other locations (i.e. via Strategies 1 to 7) but are not feasible due to a lack of funding, resources should instead be reallocated from low impact projects to these high impact projects.

Once a site is identified as an offsets donor (i.e. Strategy 8), it will often also qualify as a Strategy 9 site.

10 STRATEGY 9: MINIMUM REQUIREMENTS

Strategy aims:	To provide just the minimum amount of investment to prevent public safety risks and to avoid total sterilization of the waterway.
Qualifying criteria:	Every waterway should have a minimum amount of funding available to avert emergencies.
Site profile:	"Left to their own devices".
Examples:	Parts of the Bremmer, Oxley, Logan and Brisbane Rivers.
"Charcoal"	RGB Code: R = 50, G = 50, B = 50.

Note most of the funding drawn from these catchments is better invested elsewhere (refer to Strategies 1–7). Reasons to invest in these types of catchments however include:

10.1 Strategy 9a: Avoid public safety risks

Local governments generally have a duty to prevent harm and loss of life resulting from the following scenarios:

- Flooding caused by blocked culverts, bank destabilisation;
- Contamination of drinking water supplies;
- e-coli outbreaks – closed swimming beaches;
- Blue-green algae outbreaks and mass fish kills.

10.2 Strategy 9b: Containment of threats

There is an economic justification for the investment in this catchment where we can prevent hazards (e.g. weeds, invasive species) spilling over into neighbouring high value areas. Note these hazards need to be mobile.

10.3 Strategy 9c: Thresholds

If the waterway pollution threshold is exceeded it can lead to a sometimes-irreversible loss of value (i.e. biodiversity and species loss). We therefore need to understand the ecological thresholds of the waterway and avoid crossing this line.

10.4 Strategy 9d: Seeds of change

Small low-cost interventions can sometimes improve waterway value of their own accord with limited resources e.g. education and community based social marketing.

11 CONCLUSION

The current stormwater pollution regulations in Queensland are not leading the most effective waterway management strategy. The research behind this paper points to the need to explore different ways of managing our waterways. To this end, HLW has created the Strategic Waterways tool to assess the diverse values, hazards and needs of each waterway. From there, a unique intervention can be determined.

The main body of the paper details each of the nine possible waterway improvement strategies and outlines the scenarios where they will be most effective. This methodology allows waterway managers to construct a balanced portfolio of waterway management projects that will help to maintain, protect and enhance waterway value into the future.

ACKNOWLEDGEMENTS

Thanks, and acknowledgement must go to my colleagues at HLW especially Mr Adrian Crocetti for kindly reviewing this paper; the Department of Environment and Science for continual support of the Water by Design Program; Mr Tony Weber and Alluvium for undertaking the background scientific research for this project; and the broader stormwater industry for supporting our inquiries and providing feedback on our proposals.

REFERENCES

[1]   Weber, T. et al., *Stormwater Management Design Objectives*, Alluvium, 2018.
[2]   Browning, G.D., State Planning Policy: Expert Workshop Summary Report, Water by Design, 2020.
[3]   Browning, G.D., Let's get our ducks in a row: Novel tools for waterway prioritisation. *International Journal of Environmental Impacts*, 2(1), pp. 42–58, 2019.
[4]   Browning G.D., 50 shades of risk: A tool to analyse and prioritise complex waterway management. *World Engineering Conference*, Melbourne 2019.
[5]   Healthy Land and Water, Report Card, 2019. https://reportcard.hlw.org.au/public/media/2019-10-22/245c546b-704b-438f-a807-1068a0d33b5c/full.pdf.
[6]   DES, Environmental Protection (Water) Policy, 2009.
[7]   Water by Design State of the Streams, 2014 and 2017.
[8]   Browning, G.D., Strategic waterways. Water by Design, 2019. https://waterbydesign.com.au/news/strategic-waterways
[9]   Browning, G.D., Let's get our priorities straight. *WIT Transactions on Ecology and the Environment*, vol. 228, WIT Press: Southampton and Boston, 2018.
[10] Nasplezes, R., Davidson Street. Water by Design, 2019. https://waterbydesign.com.au/case-study/davidson-street-newmarket.
[11] Stockholm Resilience Centre, *Applying Resilience Thinking*. www.stockholmresilience.org/download/18.10119fc11455d3c557d6928/1459560241272/SRC+Applying+Resilience+final.pdf.
[12] Mullaly, J., Little McCready's Creek. Water by Design, 2019. https://waterbydesign.com.au/case-study/little-mccreadys-creek-rehabilitation-mackay-regional-council.
[13] Mullaly, J., Small Creek. Water by Design, 2019. https://waterbydesign.com.au/case-study/small-creek-ipswich-city-council.
[14] Browning, G.D., Redbank Plains. Water by Design, 2020. https://waterbydesign.com.au/case-study/redbank-plains-recreational-reserve-wetland.
[15] Port of Brisbane, Port of Brisbane website, 2020. www.portbris.com.au/Major-Projects/Offsite-Stormwater/.

INFLUENCE OF POLLUTION BUILD-UP AND PAVEMENT CROSS-SECTION ON PERMEABLE PAVEMENTS UNDER EXTREME RAINFALL EVENTS

MIRIAM FERNÁNDEZ-GONZALVO, CARMEN HERNÁNDEZ-CRESPO,
MIGUEL MARTÍN & IGNACIO ANDRÉS-DOMENECH
Instituto Universitario de Investigación de Ingeniería del Agua y del Medio Ambiente,
Universitat Politècnica de València, Spain

ABSTRACT

Permeable pavement systems (PPS) are part of the stormwater management technologies known as Sustainable Urban Drainage Systems (SUDS). The main objective of this is to analyse its buffering capacity (quantity and quality of filtered water) under extreme rainfall events. To this end, two different pavement cross-sections, complete configuration (C1) and simple configuration (C2), have been tested at laboratory scale for the 25-year return period 10-min intensity rainfall in Valencia (Spain). Four different pollutant build-up levels have been studied (at one, three, six and 12 months). For all the simulated rainfall events, the quantity (hydrographs and drained volume) and quality (DQO, TSS, TN and TP concentrations) of filtered water were characterized. On the one hand, configuration C1 reduced more than C2 the peak flow of filtered water for the same dirt and dust build-up. Moreover, peaks are wider in time, so the filtered water is better distributed. Regarding the influence of pollution build-up level, it can be observed that higher dust and dirt build-up gave rise to lower flow rates. In general, the larger the quantity of pollutant build-up accumulated on the pavement surface, the larger the volume of water was retained because it absorbed part of the rain, thus smoothing the peak flow of filtered water. On the other hand, the pollutant mass leached from the pavement C1 is lower than C2. When the amount of dust and dirt deposited was lower, the rainfall was able to leach more pollutant with respect to the total mass deposited. The results show the importance of an adequate cleaning program with a frequency no lower than once a month and a better response, in terms of water pollution retention, of the complete configuration (C1) during times of heavy rainfall.

Keywords: porous pavement, extreme rainfall intensity, pollutant build-up, infiltrated water quality, laboratory experience.

1 INTRODUCTION

Since the beginning of urban development, the water cycle had been altered. When it rains on a natural landscape, water infiltrates into the soil, evaporates, is taken by plants (evapotranspiration) and sometimes finds a way into rivers. But when it rains in an urban area, in which permeable surfaces are so scarce, a great part of rainwater turns into surface water runoff, which can cause flooding, pollution and erosion problems. Sustainable Urban Drainage Systems (SUDS) are an opportunity to prevent urban soil sealing. They can take many forms, but the main idea is to manage and use rainwater close to where it falls, i.e. at source. Therefore, the pillars of SUDS design are four: control the quantity of runoff, manage the quality of runoff to prevent pollution, create and sustain better places for people and create and sustain better places for nature [1].

Permeable Pavement Systems (PPS) are one specific type of SUDS. A large number of studies have shown the hydraulic benefits of these systems, including: reduction of surface urban runoff and peak flow attenuation, reduction of combined sewer overflows, groundwater recharge [2] and mitigation of urban heat island effect [3]. Nevertheless, there are few research about the second pillar of SUDS design (water quality), although the number has been growing in recent years [4]–[6].

WIT Transactions on Ecology and the Environment, Vol 242, © 2020 WIT Press
www.witpress.com, ISSN 1743-3541 (on-line)
doi:10.2495/WP200141

Studying the response of SUDS to high rainfall intensity events is important to demonstrate their ability to face future challenges associated with climate change, as the Intergovernmental Panel on Climate Change expects that the extreme precipitation events will become more intense and frequent [7]. It has been claimed and demonstrated that incorporating SUDS into urban planning development becomes necessary [8], and other authors also consider that urban green assemblages (equivalent definition for SUDS) has the potential to create more resilient cities with the capacity to respond to a broader scope of climate change and environmental issues [9].

The main objective of this research is to analyse the buffering capacity of permeable pavements, in terms of quantity and quality of filtered water under extreme rainfall events. The aim will be achieved quantifying the hydraulic and pollutant retention capacity of the PPS performed in laboratory under different dust and dirt build-up levels. Moreover, another goal of the study is to define the minimum cleaning frequency for PPS according to their configuration.

2 MATERIALS AND METHODS

2.1 Materials

The experimental setup designed to achieve the objectives of the study consists of a battery of laboratory scale infiltrometers, as shown in Fig. 1(a). Two configurations of pavement layers were tested: full configuration (C1) and simple configuration (C2). Both sections are also illustrated in Fig. 1(b). Configuration C2 has the same layers as C1 except the bottom layer of washed limestone aggregate 4–40 mm size. Both configurations were tested to understand the role of the gravel layer. When the traffic load is not high (e.g. on sidewalks) the bottom gravel layer is dispensable, because less structural strength is required. The concrete block characteristics, according to the information provided by the manufacturer, can be found in Hernández-Crespo et al. [10].

The rainfall simulator is based on [11] and is composed of a water storage tank, a pump and a grid of a drip irrigation pipes consisting of 7 rows with 13 drippers per row evenly distributed and placed 50 cm above the pavement surface (see Fig. 1(a)). A rain gauge is positioned below each permeable pavement to measure the infiltrated water flow, which is then collected and kept in a refrigerator until chemical analysis is performed, in less than 24 hours.

Deionized water was used (Electrical Conductivity: 10 µS/cm, pH: 7.1) to simulate rainwater. The material used to simulate pollution build-up is real dust and dirt deposited on urban roads. It was collected by the road cleaning services of the university (Universitat Politècnica de València – UPV), by a mechanical sweeper in dry conditions. The characteristics of this dust are collected in Table 1 and more detailed information is available in Hernández-Crespo et al. [10].

2.2 Methods

Rainfall simulations which reproduce the event with a return period of once in 25 years in Valencia (Spain) were performed. In particular, this rainfall intensity corresponded with 22 mm in 10 min (132 mm/h). A plastic mesh was included in the setup installation, between the rain simulator and the chamber that contains the paver profile, according to Naves et al. [12] to break rainfall drops reducing its size and improving its spatial distribution.

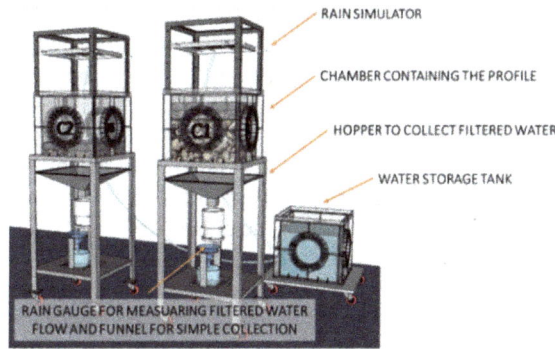

(a)

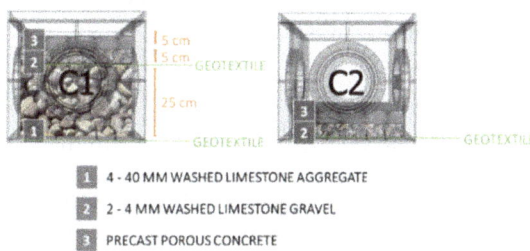

1 4 - 40 MM WASHED LIMESTONE AGGREGATE

2 2 - 4 MM WASHED LIMESTONE GRAVEL

3 PRECAST POROUS CONCRETE

(b)

Figure 1:   (a) Scheme of experimental setup; and (b) Cross-section configuration C1 (left) and configuration C2 (right).

Table 1:   Physical-chemical characteristics of the dust and dirt used in the tests. Results are expressed on a dry weight (dw) basis.

Physico-chemical variable		Average value
Moisture (% dw)		2.02
Organic matter (% dw)		8.23
Organic carbon (% dw)		1.83
Total nitrogen (mg/kg dw)		1351.0
Total Phosphorus (mg/kg dw)		302.5
Electrical conductivity$_{1:5}$ (μS/cm)		890
pH$_{1:5}$		8.0
Particle size distribution (%dw)	Gravel (>2 mm)	13.2
	Sand (0.063–2 mm)	85.0
	Silt and clay (<0.063 mm)	1.8

The influence of pollution build-up was studied by dry sprinkling sediments on the pavement surface. The applied deposition rate (5 g/m^2/d) was selected according to the information provided by the cleaning services (mass collected, surface swept and cleaning frequency) and checked against scientific references [13], [14]. Four different pollutant build-up levels effects have been studied: one month (140 g/m^2), three months (420 g/m^2), 6 months (840 g/m^2) and 1 year (1825 g/m^2).

Only three infiltrometers were available for testing, one with configuration C1 and two with configuration C2 (Replica 1 and Replica 2). In general, the methodology followed for the trials was applying progressively the dust (1, 3, 6 and 12 months) and simulating the rainfall in each case. Three weeks have passed between tests, on the same infiltrometer, to ensure the same initial moisture conditions. There was an exception for configuration C2, in which 1 month and 6 month of pollutant build-up level were performed and after that, the surface was cleaned with a vacuum cleaner and a brush, and the assays of 3 and 12 months of build-up levels were performed to complete the data series.

The infiltrated water was collected and kept in the refrigerator until analysis within twenty-four hours. The water samples were analysed for the following parameters: Total Nitrogen (TN) (Spectroquant® test: [15] photometry), Total Phosphorus (TP) (Spectroquant® test: digestion [16]), Chemical Oxygen Demand – COD (Spectroquant® test: [17]), Total and Volatile Suspended Solids [18], [19] and turbidity with a TN100 Eutech turbidity-meter. WTW-Multi 340i probes were used to measure: temperature, dissolved oxygen (CellOx® 325), conductivity (TetraCon®) and pH (SenTix®41).

3 RESULTS

3.1 Influence of pollution build-up and rainfall intensity on infiltrated water quantity

The hydrographs of infiltrated water obtained in the different tests performed are shown in Fig. 2. Additionally, drained volumes are represented through the cumulative volume curves. Fig. 2(a) collects the results for 1-month pollutant build-up, (b) for 3 months, (c) for 6 months and (d) for 12 months. For 1 and 6 months of dust and dirt accumulation, the maximum flow rates and the lower lag times were reached for simple configuration (Fig. 2(a) and (c)). Lag time is defined as the time elapsed between the beginning of the rainfall and the beginning of the outflow from permeable pavement, as defined in Rodriguez-Hernandez et al. [20]. For 3- and 12-month build-up, these variables did not show a clear pattern.

For configuration C1, the larger the quantity of pollutant accumulated on the pavement surface, the lower flow rates and the longer lag time. This happened during the first three build-up simulations (1, 3 and 6 months of dirt and dust accumulated) where the flow rate values were 1.36, 1.34 and 1.33 $l/m^2/min$, respectively, and the lag times increased from 7 to 9 (Fig. 2(a)–(c)). It seems that dust and dirt accumulated on the pavement surface absorbed part of the rain, thus smoothing the peak flow of filtered water and delaying the hydrographs. For the fourth test (12 months), peak flow increased (until 1.75 $l/m^2/min$) and lag time was 5 minutes; thus giving a peak flow higher than the assay performed under 1-month build-up and a lower lag time (Fig. 2(d)). Too much dust (1825 g/m^2) possibly clogged the pores of the pavement, and rainwater could have found preferential paths, short-circuiting and, thus, reducing its buffering capacity. In this assay, runoff was formed, proving some clogging.

For both replicas performed on simple configuration (C2), it can be also observed, for 1 and 6 months, a higher quantity of build-up produces a reduction of peak flows and an increase of lag time (Fig. 2(a) and (c)). However, 3 and 12-months tests (Fig. 2(b) and (d)) produced lower peak flows (1.2 $l/m^2/min$ and 0.7 $l/m^2/min$, respectively) and significantly higher lag times (6 and 7 minutes, respectively).

As it has been stated in previous section, infiltrometers with configurations C2 were tested with one and six months of dust and dirt build-up and then were cleaned with a domestic vacuum cleaner and a brush. After that, three- and 12-months tests were performed. When the cleaning works finished, it was observed that some dirty remained between the joints of the paving blocks. This situation might have caused a delay in rain infiltration, leading to the

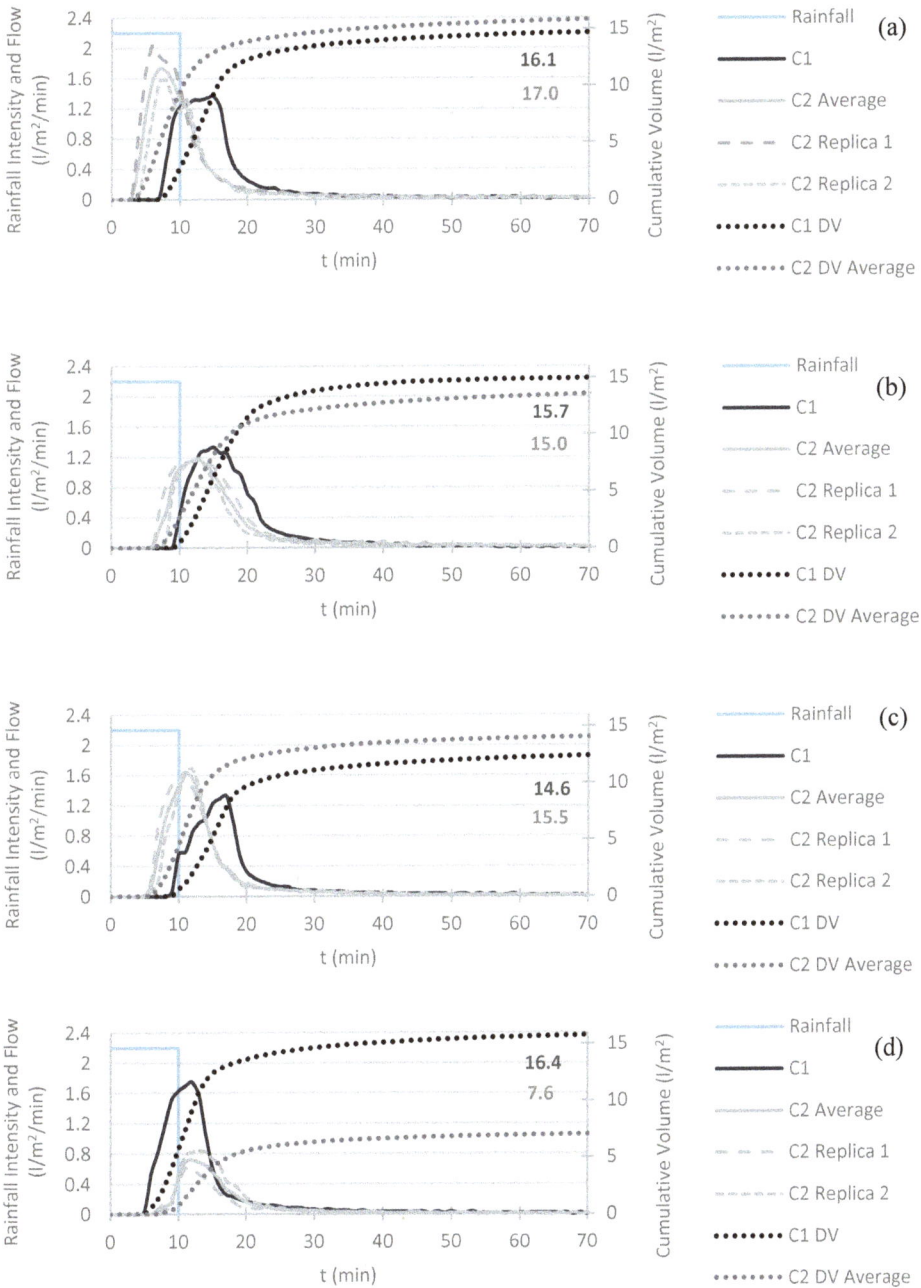

Figure 2:   Hydrographs of infiltrated water under different pollution build-up levels for the tested configurations C1 and C2. (a) 1 month; (b) 3 months; (c) 6 months; and (d) 12 months. Cumulative volume curves (C1 DV and C2 DV average) are displayed on the secondary axis and numerical data indicate the total drained volume (average value for configuration C2).

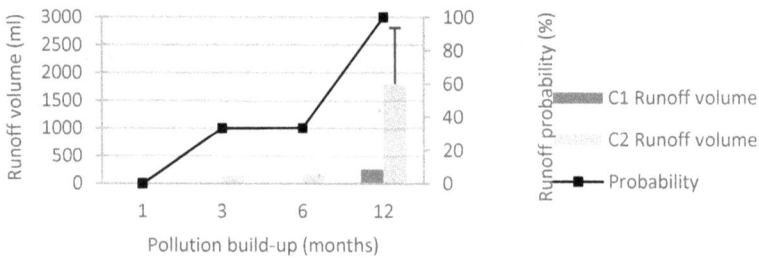

Figure 3:   Runoff volume in millilitres for 1, 3, 6 and 12 months of pollutant build-up and both tested configuration (C1 and C2). The probability of runoff occurrence, according to the number of tests carried out, are displayed on the secondary axis.

runoff production (140.5 ml for three months of build-up and 2516 ml for one year). Most researchers in the field claim pressure/power washing with water and vacuum sweeping (or a combination of these) are the most recommended methods to rehabilitate clogged permeable concrete [21], [22].

In terms of total drained or retained volume, permeable pavements with both configurations were always able to retain a significant part of the total rainfall volume applied, at least 22% (one-month test – C2). In short, retention varies between 24%–34% for complete configuration and between 22%–33% in the simple one (C2).

About runoff generation, configuration C2 produced a runoff volume that increased as the pollution build-up level did (Fig. 3). In contrast, configuration C1 only produced runoff for 1 year of build-up level (263 ml). Right vertical axis in Fig. 3 shows the probability of runoff occurrence observed in the performed tests. For one month of dust and dirt accumulation, runoff does not occur. Whereas, for three and six months only one of the replicas performed in the simple configuration (C2) produced it. Finally, one-year pollutant build-up produced runoff volume in all the tests. In Drake and Bradford [23] recommended frequency of maintenance ranges from at least once per year and [24], [25] to two to four times per year, depending on the site and weather conditions [26]. Taking into account both configuration, the results show that it is necessary to clean the PPS with a frequency no lower than once a month to maintain an adequate permeability rate and its capability to prevent surface runoff. It could be a good idea, a basic monthly cleaning with mechanical sweeper, just like in any other area of the city.

3.2 Influence of pollution build-up and rainfall intensity on the quality of infiltrated water

This section presents the results of the water quality, for C1 and C2 configurations. Fig. 4 shows pollutant concentrations for filtered water through the pavements and the runoff generated if it is the case.

Regarding the quality of filtered water, configuration C1 reached lower concentrations of COD, TN and TP than configuration C2 (Fig. 4(a), (c) and (d) respectively). COD increased progressively with pollutant build-up, reaching concentrations between 77 mg/l (Configuration C1) and 174 mg/l (Configuration C2) for one year of dust accumulation (value above 125 mg/l; discharge limit of urban wastewater treatment plants in Spain). This rise is more pronounced for configuration C2 (Fig. 4(a)).

(a) COD

(b) TSS

(c) TN

(d) TP

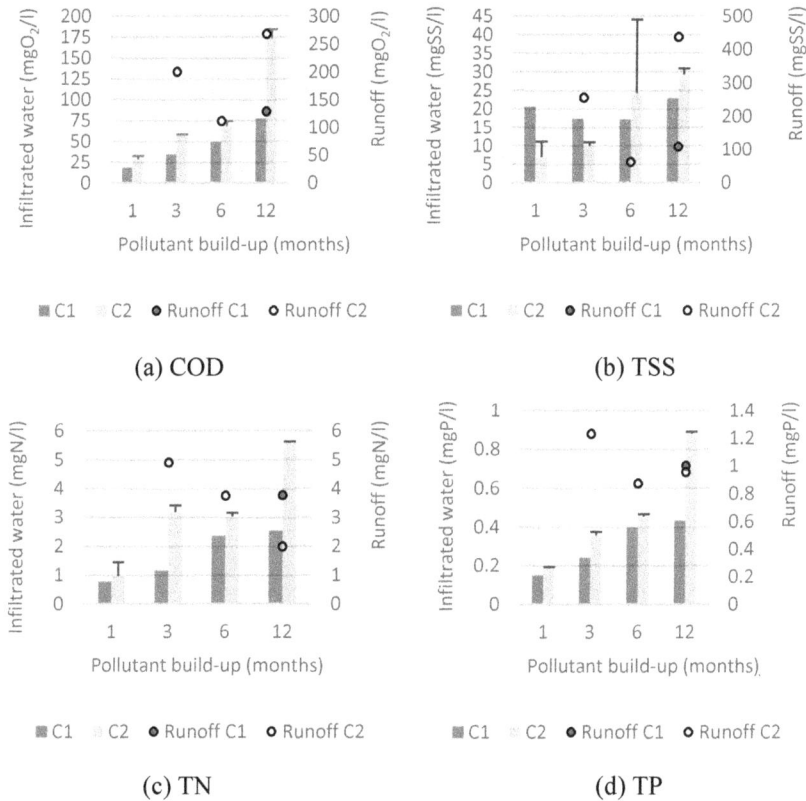

Figure 4: (a) Chemical oxygen demand (COD); (b) Total Suspended Solids (TSS); (c) Total Nitrogen (TN); and (d) Total Phosphorus (TP) concentrations in the different assays performed (1, 3, 6 and 12 months of pollutant build-up in configuration C1 and C2). Vertical left axis refers to infiltrated water concentrations and vertical right axis to runoff concentrations. The standard deviation is only shown for infiltrated water.

Total Nitrogen and Total Phosphorus concentrations in complete configuration tests have a tendency to increase more in the first tests (the difference between 1 and 3 months of build-up levels is higher) and to steady with dust accumulation (results for 6 and 12 months are quite similar). By contrast, concentrations of these pollutants in simple configuration increase progressively with build-up (Fig. 4(c) and (d)).

The behaviour of TSS concentrations (Fig. 4(b)) is different for both configurations. C2 TSS concentrations increase gradually with deposition. The larger the quantity of pollutant build-up, the larger the concentration of infiltrated water become. Conversely, there is not an important difference among TSS concentration in the water drained by pavements with complete configuration for the different levels of dirty accumulation. Furthermore, it should be highlighted that the amount of TSS leached through the PPS is low in comparison with the amount deposited on the surface, as it is discussed below.

About runoff, as it is commented in the previous section, configuration C2 produced for 3, 6 and 12 months of pollutant build-up, and configuration C1 only generated for 1-year build-up level. More pollutant accumulation produces higher runoff volumes. For configuration C2, three-months runoff volume was more concentrated than six-months one. One-year tests produced the higher runoff volumes (Fig. 3) and, especially for configuration C2, very high pollutant concentrations (268 mgO$_2$/l of COD, 436.8 mgSS/l of TSS, 1.98 mgN/l of TN and 0.955 mgP/l of TP).

So far, the results obtained have been analysed for the concentration of pollutants, but it is also important to analyse the results in terms of released mass load. Fig. 5 shows the percentage of mass leached within the infiltrated water with respect to the total mass present

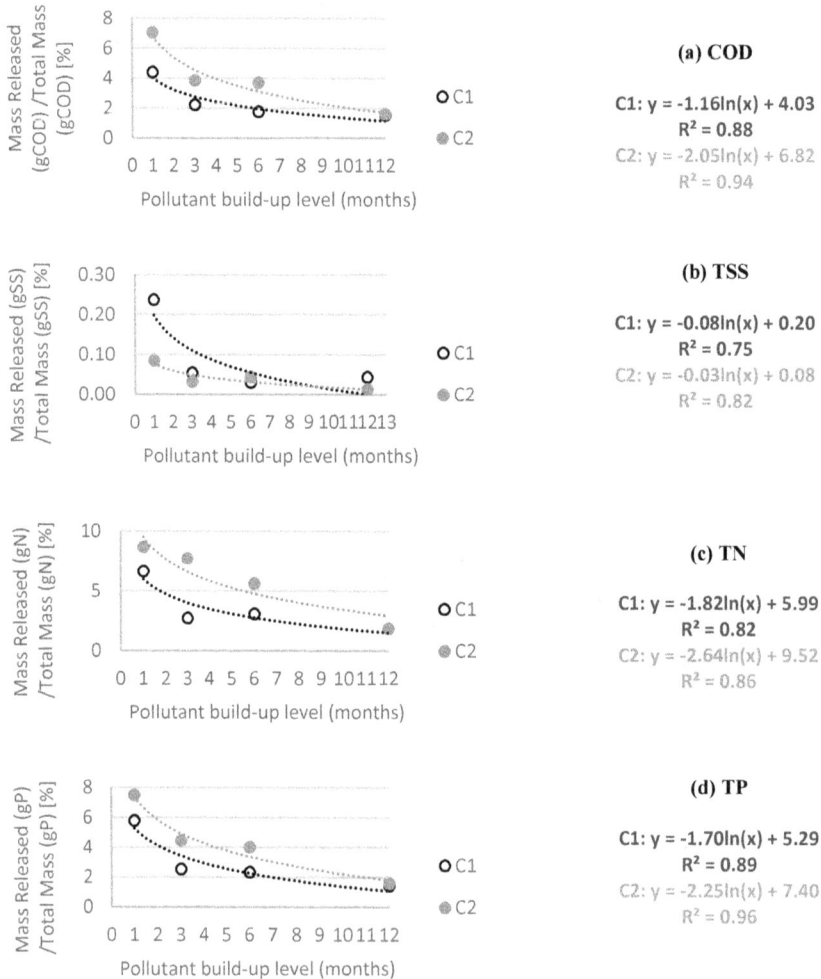

Figure 5: Evolution of mass of pollutant released with respect to the total mass present in the dust and dirt deposited on the pavement surface (%) with pollutant build-up level. Dots represent the data (average data for configuration C2) and dotted curves a logarithmic adjustment. On the right are the equations of the adjustment for each configuration (C1 and C2) and the coefficient of determination (R2).

in the dust and dirt accumulated on the pavement surface. It is important to carry out the analysis in terms of mass to quantify the pollutant retention capacity of the different PPS configurations. This analysis reaffirms the higher retention capacity of configuration C1 (especially for lower values of pollution build-up) considering the effect of volume and concentration together.

The percentage of TSS leached is the lower one, compared with the rest of the variables studied, and the differences between build-up levels are also very low. Both configurations have the greatest retention capacity for the TSS, for all the build-up levels tested. The quantity of solids released is so small for both configurations, thus the difference obtained can be considered negligible.

For all the variables, this percentage of mass leached within the filtered water, can be adjusted to a logarithmic model. That means the mass of pollutant leached trends to reach a constant value with a certain mass of dust and dirt deposited on the pavement surface. This information is interesting and highlight the potential of permeable pavements despite its lack of maintenance. However, as it was stayed above, it is highly recommendable to perform an adequate cleaning programme, to prevent the production of runoff and its potential discharge to aquatic environment.

4 CONCLUSIONS

Results have shown that PPS have an important improvement potential compared with conventional pavements. These systems have the capacity to reduce the peak flow and the volume of drained water, temporarily dampening the rain hydrograph. Moreover, they can reduce or even completely avoid the production of runoff in the extreme rainfall events tested and can decrease the pollutant concentrations respect to the surface runoff. However, PPS need to be maintained correctly to perform these advantages and this maintenance must be more frequent for simple configurations.

During extreme events, complete configuration has presented higher buffering capacity of the rainfall events than simple configuration. Peak flows, drained volumes and pollutant concentrations of infiltrated water were higher for configuration C2, impacting more on receiving environment. Besides, the runoff occurrence probability is higher for simple configuration. Only when its surface has not been cleaned in a year, complete configuration produces runoff, which volume will be probably less than 1.2 l/m^2. Elseways, simple configuration produces runoff for pollution build-up levels higher than one month.

Overall, both configurations could be used in areas where heavy rainfalls occur. Nevertheless, the benefits of complete configuration could make it useful than simple configuration, although its higher structural strength is not required. A cost-benefit analysis should be carried out before making the final decision. For both configurations, it would be advisable to clean the surface of the PPS once a month to avoid surface runoff production and less leachability of pollutants.

ACKNOWLEDGEMENTS

This research has been developed within the SUPRIS-SUPel project framework (Ref. BIA2015-6540-C2-2-R MINECO/FEDER, EU). This project had been financed by the Ministry of Economy and Competitiveness of Spain through the General State Budget and the European Regional Development Fund (ERDF). Currently, the second phase of the project, ENGODRAIN – RTI2018-094217-B-C31, financed by the State Agency for Research from Spain through the General State Budget and the European Regional Development Fund (ERDF), is continuing. The authors would also like to thank the companies Quadro Concrete Prefabs S.L., TenCate Geosynthetics Iberia S.L., Atlantis SUDS

S.L. and SECOPSA for supplying materials for the tests. Finally, the hiring of Miriam Inmaculada Fernández Gonzalvo has been possible to the program of subsides for hiring of pre-doctoral research personnel (ACIF/2018/111) of the Generalitat Valenciana.

REFERENCES

[1]   Woods Ballard, B. et al., *The SuDS Manual*, CIRIA: London, 2015. ISBN: 978-0-86017-760-9.

[2]   Tedoldi, D., Chebbo, G., Pierlot, D., Kovacks, Y. & Gromaire, M.C., Impact of runoff infiltration on contaminant accumulation and transport in the soil/filter media of sustainable urban drainage systems: a literature review. *Science of the Total Environment*, **569–570**, pp. 904–926, 2016.

[3]   Liu, Y., Li, T. & Peng, H., A new structure of permeable pavement for mitigating urban heat island. *Science of the Total Environment*, **634**, pp. 1119–1125, 2018. https://doi.org/10.1016/j.scitotenv.2018.04.041.

[4]   Brown, R.A. & Borst, M., Nutrient infiltrate concentrations from three permeable pavement types. *Journal of Environmental Management*, **164**, pp. 74–85, 2015. https://doi.org/10.1016/j.jenvman.2015.08.038.

[5]   Drake, J., Bradford, A. & Van Seters, T., Stormwater quality of spring-summer-fall effluent from three partial-infiltration permeable pavement systems and conventional asphalt pavement. *Journal of Environmental Management*, **139**, pp. 69–79, 2014. https://doi.org/10.1016/j.jenvman.2013.11.056.

[6]   Kamali, M., Delkash, M., Tajrishy, M., Evaluation of permeable pavement responses to urban surface runoff. *Journal of Environmental Management*. 187, pp. 43–53, 2017. https://doi.org/10.1016/j.jenvman.2016.11.027.

[7]   IPCC, Summary for policymakers. *Global Warming of 1.5°C. An IPCC Special Report on the Impacts of Global Warming of 1.5°C above Pre-Industrial Levels and Related Global Greenhouse Gas Emission Pathways, in the Context of Strengthening the Global Response to the Threat of Climate Change, Sustainable Development, and Efforts to Eradicate Poverty*, eds V. Masson-Delmotte, et al., World Meteorological Organization: Geneva, 2018.

[8]   Jenkins, K., Surminski, S., Hall, J. & Crick, F., Assessing surface water flood risk and management strategies under future climate change: Insights from an Agent-Based Model. *Science of the Total Environment*, **595**, pp. 159–168, 2017. http://dx.doi.org/10.1016/j.scitotenv.2017.03.242.

[9]   Baron, N. & Petersen, L.K., Understanding controversies in urban climate change adaption. A case study of the role of homeowners in the process of climate change adaptation in Copenhagen. *Nordic Journal of Science and Technology Studies*, **3**(2), pp. 4–13, 2015. https://doi.org/10.5324/njsts.v3i2.2159.

[10]   Hernández-Crespo, C., Fernández-Gonzalvo, M., Martín, M. & Andrés-Doménech, I., Influence of rainfall intensity and pollution build-up levels on water quality and quantity response of permeable pavements. *Journal Science of Total Environment*, **684**, pp. 303–313, 2019. https://doi.org/10.1016/j.scitotenv.2019.05.271.

[11]   Rodriguez-Hernández, J., Andrés-Valeri, V.C., Ascorbe-Salcedo, A. & Castro-Fresno, D., Laboratory study on the stormwater retention and runoff attenuation capacity of four permeable pavements. *Journal of Environmental Engineering*, **142**(2), pp. 04015068, 2016.

[12]   Naves, J., Puertas, J., Anta, J., Suárez, J. & Regueiro-Picallo M., Montaje y calibración de un simulador de lluvia para estudios de drenaje urbano. *V Jornadas de Ingeniería del Agua*, 24–26 de Octubre, A Coruña, 2017.

[13]   Wijerisi, B., Egodawatta, P., McGree, J. & Goonetilleke, A., Process variability of pollutant build-up on urban road surfaces. *Science of the Total Environment*, **518–519**, pp. 434–440, 2015. https://doi.org/10.1016/j.scitotenv.2015.03.014.

[14]   Zhao, H. et al., Is the wash-off process of road-deposited sediment source limited or transport limited? *Science of the Total Environment*, **563–564**, pp. 62–70, 2016. https://doi.org/10.1016/j.scitotenv.2016.04.123.

[15]   ISO 11905-1:1997, Water Quality Determination of nitrogen. Part 1: method using oxidative digestion with peroxodisulfate.

[16]   ISO 6878-1:1986, Water Quality. Determination of phosphorus. Part 1: ammonium molybdate spectrometric method.

[17]   ISO 15705, 2002, Water Quality. Determination of the Chemical Oxygen Demand Index (ST-COD). Small-scale Sealed-tube Method.

[18]   UNE 77034:2002, Water Quality. Determination of volatile and fixed suspension solid.

[19]   UNE-EN 872:2006, Water Quality. Determination of suspended solids. Method by filtration through glass fibre filters.

[20]   Rodriguez-Hernandez, J., Andrés-Valeri, V.C., Ascorbe-Saceldo, A. & Castro-Fresno, D., Laboratory study on the stormwater retention and runoff attenuation capacity of four permeable pavements. *Journal of Environmental Engineering*, **142**(2), pp. 04015068, 2015.

[21]   Environmental Protection Agency (EPA). National Menu for Best Management Practices, Post-construction Stormwater Management. Environmental Protection Agency (EPA), 2004.

[22]   Golroo, A. & Tighe, S.L., Pervious concrete pavement performance modelling: an empirical approach in cold climates. *Canadian Journal of Civil Engineering*, **39**(10), pp. 1100–1112, 2012.

[23]   Drake, J.A.P. & Bradford, A., Assessing the potential for restoration of surface permeability for permeable pavements through maintenance. *Water Science and Technology*, **68**(9), pp. 1950–1958, 2013.

[24]   Gunderson, J., Pervious Pavements: New Findings about Their Functionality and Performance in Cold Climates. Stormwater Magazine, Sep. 2008.

[25]   Henderson, V. & Tighe, S., Evaluation of pervious concrete pavement performance in cold weather climates. *International Journal of Pavement Engineering*, **13**(3), pp. 197–208, 2012.

[26]   Kia, A., Wong, Hong S. & Cheeseman, C.R., Clogging in permeable concrete: A review. *Journal of Environmental Management*, **193**, pp. 221–233, 2017.

Author index

WIT*PRESS* ...for scientists by scientists

Urban Water Systems & Floods III

Edited by: **S. MAMBRETTI**, *Polytechnic of Milan, Italy and* **D. PROVERBS**, *Birmingham City University, UK*

Flooding is a global phenomenon that claims numerous lives worldwide each year. Apart from the physical damage to buildings, contents and loss of life, which are the most obvious, impacts of floods upon households and other more indirect losses are often overlooked. These indirect and intangible impacts are generally associated with disruption to normal life and longer term health issues. Flooding represents a major barrier to the alleviation of poverty in many parts of the developing world, where vulnerable communities are often exposed to sudden and life-threatening events.

As our cities continue to expand, their urban infrastructures need to be re-evaluated and adapted to new requirements related to the increase in population and the growing areas under urbanization. Topics such as contamination and pollution discharges in urban water bodies, as well as the monitoring of water recycling systems are currently receiving a great deal of attention from researchers and professional engineers working in the water industry. The papers contained in this volume cover these problems and deals with two main urban water topics: water supply networks and urban drainage.

Originating from the 7th International Conference on Flood and Urban Water Management, the included research works include innovative solutions that can help bring about multiple benefits toward achieving integrated flood risk and urban water management strategies and policy.

ISBN: 978-1-78466-379-7   eISBN: 978-1-78466-380-3
Published 2020 / 196pp

www.ingramcontent.com/pod-product-compliance
Lightning Source LLC
Chambersburg PA
CBHW080243230326
41458CB00097B/2989